ALS
Advances in Life Sciences

Progress in Membrane Biotechnology

Edited by
J.C. Gomez-Fernandez
D. Chapman
L. Packer

Birkhäuser Verlag
Basel · Boston · Berlin

Editors' addresses:

Prof. Juan C. Gomez-Fernandez
Departamento de Bioquimica
y Biologia Molecular
Facultad de Veterinaria
Universidad de Murcia
E-30071 Murcia/Spain

Prof. Dennis Chapman
University of London
Royal Free Hospital School of Medicine
Dept. of Protein & Molecular Biology
Rowland Hill Street
London NW3 2PF/UK

Prof. Lester Packer
Molecular & Cell Biology Dept.
University of California
251, Life Science Addition
Berkeley, California 94720/USA

Deutsche Bibliothek Cataloging-in-Publication Data

Progress in membrane biotechnology / ed. by J.C. Gomez-
Fernandez ... - Basel ; Boston ; Berlin : Birkhäuser, 1991
  (Advances in life science)
   ISBN 3-7643-2666-2 (Basel ...)
   ISBN 0-8176-2666-2 (Boston)
NE: Gomez-Fernandez, Juan C. [Hrsg.]

© 1991 Birkhäuser Verlag
     P.O. Box 133
     4010 Basel
     Switzerland

Printed from the authors' camera-ready manuscripts
on acid-free paper in Germany
ISBN 3-7643-2666-2
ISBN 0-8176-2666-2

Contents

v

Preface

It is well known that basic science can trigger an invention of considerable technological and commercial importance. Indeed basic science and invention are often inextricably linked, each being able to catalyse the other. To engender such developments it is important that there should be good communication between the scientist and the technologist. The field of membrane biotechnology is a growing field where such communication is increasingly taking place and where new inventions are occurring.

This book provides an overview of this developing field. It contains chapters by scientist and technologists working in the field of Membrane Biotechnology. The chapters cover the latest advances in basic science as well as some recent technological applications. The basic topics include the application of dynamic X-ray diffraction to lipid water systems, FTIR spectroscopy applied to membrane proteins, fluorescent analogues of phosphoinositides, studies of platelet activating factor, antibody binding to model membranes and phospholipase C induced fusion.

The technological topics described include the development of new haemocompatible materials based upon biomembrane mimicry, new lung surfactant materials, drug delivery systems including liposomes and the development of new biosensors including Langmuir Blodgett films.

The meeting showed that there are many other useful applications in the pipeline. The potential for new polymeric drug delivery systems, of ion selective systems based on the knowledge of ion-channel protein structures, of new plastics for cell growth and cellular engineering for artificial organs. These are among the interesting developments that are emerging in this field.

Dennis Chapman
Juan Carmelo Gomez-Fernandez

Our thanks are due to financial assistance from:

- International Union of Pure and Applied Biophysics

- International Union of Biochemistry

- European Research Office, U.S. Army (London)

- Office of Naval Research (London)

- Dirección General de Investigación Cientifica y Tecnológica (Madrid).

- Comisión Interministerial de Ciencia y Tecnologia; Programa Nacional de Biotecnologia (Madrid)

- Instituto de Fomento de la Región de Murcia

- Dirección Regional de Educación y Universidad (Murcia)

One of the Editors (D. Chapman) would like to thank the Banco Bilbao y Vizkaya Foundation for a fellowship which enabled the preparation of this volume.

Progress in Membrane Biotechnology
Gomez-Fernandez/Chapman/Packer (eds.)
© 1991 Birkhäuser Verlag Basel/Switzerland

BASIC AND TECHNOLOGICAL STUDIES OF BIOMEMBRANE COMPONENTS

D. Chapman and P.I. Haris

Department of Protein & Molecular Biology, Royal Free Hospital School of
Medicine, Rowland Hill Street, London NW3 2PF

SUMMARY:

As part of our basic studies we have employed various biophysical techniques
for the study of biomembrane structure.   Fourier transform infrared (FTIR)
spectroscopy has been used to probe the structure of membrane associated
polypeptides and proteins.   Here we discuss the recent development of methods
for quantitative analysis of protein secondary structure for both soluble and
membrane proteins.   With regard to technological studies, we are pursuing
research to develop new haemocompatible biomaterials. Results of some our
recent studies are presented.

INTRODUCTION

We have continued our studies of biomembrane systems using a range of
physical techniques (Chapman & Benga, 1984; Lee *et al.*, 1985; Haris *et al.*,
1989a; Mitchell *et al.*, 1988; Villalain *et al.*, 1989; Haris *et al.*, 1989b).   Our
recent basic studies are particularly aimed at examining polypeptides and proteins
associated with biomembranes. Thus we have been developing Fourier transform
infrared (FTIR) spectroscopy for qualitative and quantitative studies of such
systems and applying this technique to membrane proteins, signal polypeptides
and ion-channel proteins and polypeptides.   We are also supporting these studies
with additional techniques such as circular dichroism (CD) spectroscopy and
nuclear magnetic resonance (NMR) spectroscopy.

In addition to these basic studies we have also continued to develop our technological work associated with new biomaterials with particular interest for the development of haemocompatible biomaterials.

In this presentation we will review briefly some of our basic and our technological studies.

BASIC STUDIES

Quantitative analysis of protein structure using FTIR spectroscopy: FTIR studies of protein and polypeptide systems in aqueous media (H_2O and ^2H_2O) are now well documented. Considerable use has been made of the amide I band in the 1700-1600cm^{-1} region of the spectrum to obtain qualitative information about the secondary structures which are present in these systems. The use of deconvolution and second derivative computing methods producing band narrowing enables overlapping bands hidden in the amide I band envelope to be distinguished. We have now studied a range of polypeptides, proteins and membrane proteins (Lee *et al.*, 1985; Haris *et al.*, 1989; Mitchell *et al.*, 1988; Villalain *et al.*, 1989;

Table I: Protein studies using FTIR spectroscopy

	Reference
Membrane proteins	
Bacteriorhodopsin	Lee *et al.*, 1985
Rhodopsin	Haris *et al.*, 1989a
H^+/K^+-ATPase	Mitchell *et al.*, 1988
Ca^{++}-ATPase	Villalain *et al.*, 1989
Cytochrome Oxidase	Haris *et al.*, 1989b
Soluble proteins	
α-Anti-trypsin	Haris *et al.*, 1990
Concanvalin A	Alvarez *et al.*, 1987
Phospholipase A2	Kennedy *et al.*, 1990
Factor H	Perkins *et al.*, 1989
Properdin	Perkins *et al.*, 1988
Ribonuclease	Haris *et al.*, 1986

Haris *et al.*, 1989; Haris *et al.*, 1990; Alvares *et al.*, 1987; Kennedy *et al.*, 1990; Perkins *et al.*, 1988; Perkins *et al.*, 1989; Haris *et al.*, 1986). Some of these are listed in Table I.

To provide further qualitative information on the structure of polypeptides and proteins we have also examined a series of polypeptides which are known to possess a 3_{10}-helical structure. We have demonstrated that a high amide I frequency ($\sim$1662-1666 cm^{-1}) occurs with the presence of a 3_{10}-helical structure. This is consistent with the weaker hydrogen bonding present as compared with the normal α-helix (Kennedy *et al.*, 1991).

New quantitative FTIR studies of the secondary structures of polypeptides and proteins have also been developed in our laboratory to overcome some of the difficulties present in previous methods. The method of Lee *et al.*, (1989) and Lee *et al.*, (1990) applies factor analysis and multiple linear regression to aqueous FTIR spectra of 18 proteins with known secondary structures. The best results for α-helix, β-sheet, and turn structure were obtained by this method when the amide I region from normalized FTIR spectra was used to construct the calibration set. Mahalonobis' statistic was also introduced to evaluate whether an unknown lies within the vector space of the basis set. In this case the X-ray structure information was based upon the work of Levitt & Greer, (1977).

In a very recent investigation a CD matrix method has been used for (Sarver *et al.*, 1991) the analysis of protein secondary structure with FTIR spectra replacing the CD spectra. This method applies singular value decomposition techniques to determine the mathematical relationship between the X-ray secondary structures of known proteins and the ir spectra of the known proteins in the amide I region.

Further quantitative studies of water soluble proteins and membrane proteins have also been made in our laboratory (Lee *et al.*, 1991). The new method (a partial least squares analysis) differs from the earlier method mainly in the calculation of the factors. Some predictions for the secondary structures of some membrane proteins using the 1700-1600 cm^{-1} region using a calibration set containing 17 soluble proteins is shown in Table II.

Table II: Secondary structure prediction for membrane proteins

Membrane Proteins	% α-helix	% ß-sheet	% turn
Ca^2-ATPase	65.1	15.7	6.8
H$^+$/K$^+$-ATPase	47.6	33.0	21.5
Photosystem II Reaction Centre	60.0	17.6	22.2
Bacteriorhodopsin	71.5	5.8	-17.2

The results obtained for the membrane proteins are consistent with previous estimations based on predictions from sequence data, electron diffraction data or qualitative interpretation of infrared spectra. For example, the prediction for Ca^{2+}-ATPase is consistent with earlier measurements by Raman, infrared and CD spectroscopies and a prediction based on amino-acid sequence. In addition, the helix estimation for bacteriorhodopsin is in close agreement with the available diffraction data and sequence prediction. (The appearance of a negative value is probably a reflection of the spectral properties of bacteriorhodopsin lying outside the range of properties defined by the calibration set). Nevertheless this approach, which is free from artefacts caused by the membrane environment and may readily be applied to membrane proteins in their native form, shows high promise for accurate quantitative descriptions of secondary structure.

We have also been applying these spectroscopic techniques to studying signal polypeptides and ion channel protein systems.

Signal polypeptides: The role of signal sequences involved in protein secretion have been extensively studied. However, the detailed molecular mechanisms of their function is still poorly understood. The structure of signal peptides is thought to be important for the protein translocation process. We have been involved in the investigation of structures of a number of signal peptides in our laboratory. Recent unpublished studies in our laboratory with a mitochondrial signal polypeptide has shown that it adopts a predominantly α-helical structure in a membrane environment. The structural studies were conducted using FTIR, CD and NMR spectroscopy.

Ion channel polypeptides: Studies in our laboratory are also being conducted to determine the structure of membrane spanning ion channel proteins. We are presently looking at the structure using FTIR, CD and NMR spectroscopy of polypeptides which we have synthesised corresponding to segments of a K^+ ion channel protein from Drosophila. These correspond to the S2, S4 and H5 segments. A recent study has suggested that the H5 segment corresponds to a sequence of 17 amino acids which form a β-hairpin that extends into the membrane somewhat analogous to a porin type structure (Yool and Schwarz, 1991). The H5 region is thought to line the pore of the channel.

TECHNOLOGICAL STUDIES

Studies of lipid polar surfaces in blood: In earlier reports we have pointed to the high predominance of phosphoryl-choline group (PC-groups - 90%) which occur on the outer lipid layer of red blood cells and platelet cells. These PC-groups arise from the presence of the lecithin and sphingo-myelin molecules on the outer surface. We have used this fact to develop new haemocompatible biomaterials. Our first studies used polymerizable lipids (Johnston et al., 1980) and later developments include grafting new polymer synthesis. This work has led us to consider the interaction of other lipid polar surfaces with blood. We have therefore extended our studies to include negatively charged, positively charged as well as including further studies of zwitterionic PC-lipids.

We have used the technique of material thrombelastography (MTEG). This involves coating a metal pin and cuvette and examining clot formation. We have also carried out various other coagulation tests to investigate the effect of the different polar lipid surfaces when present in blood.

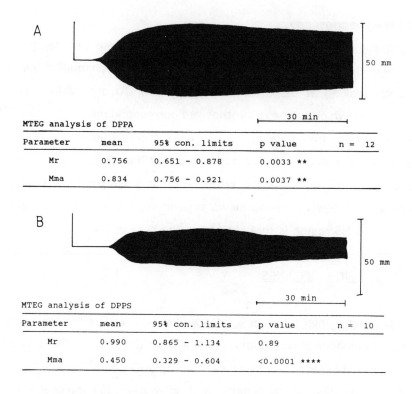

A

50 mm

|← 30 min →|

MTEG analysis of DPPA

Parameter	mean	95% con. limits	p value	n = 12
Mr	0.756	0.651 - 0.878	0.0033 **	
Mma	0.834	0.756 - 0.921	0.0037 **	

B

50 mm

|← 30 min →|

MTEG analysis of DPPS

Parameter	mean	95% con. limits	p value	n = 10
Mr	0.990	0.865 - 1.134	0.89	
Mma	0.450	0.329 - 0.604	<0.0001 ****	

Fig. 1:   MTEG analysis of negatively charged phospholipids. Typical traces are shown for (A) dipalmitoylphosphatidic acid - DPPA, (B) dipalmitoylphosphatidylserine - DPPS. A thin film of lipid was coated onto the surface of the piston and cuvette by dipcoating, and its effect on whole blood coagulation measured simultaneously with the untreated surface.

The MTEG technique is a simple technique which provides information about whole blood coagulation (Bird *et al.*, 1988). We also measure the coagulation time of citrated plasma in the presence of multilamellar liposomes using the Stypven tests. Clotting Factor Assays were used to investigate specific areas of the coagulation cascade.

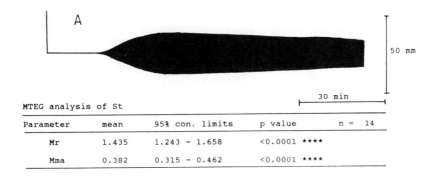

MTEG analysis of St

Parameter	mean	95% con. limits	p value	n = 14
Mr	1.435	1.243 - 1.658	<0.0001 ****	
Mma	0.382	0.315 - 0.462	<0.0001 ****	

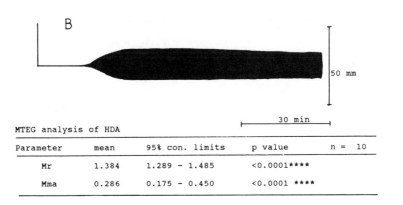

MTEG analysis of HDA

Parameter	mean	95% con. limits	p value	n = 10
Mr	1.384	1.289 - 1.485	<0.0001****	
Mma	0.286	0.175 - 0.450	<0.0001 ****	

Fig. 2:   MTEG analysis of positively charged surfaces. typical traces are shown for (A) Stearylamine - St and (B) hexaderylamine - HDA.

The results using the MTEG method show the dramatic differences which exists between negative, positive and the neutral zwitterion surfaces in their interaction with blood (See Fig. 1-3)(Hall *et al.*,).   MTEG analysis in the presence of negatively charged lipids (particularly dipalmitoylphospha- tidic acid - DPPA, dipalmitoylphosphatidylserine - DPPS and dimyristoylphosphatidyl-glycerol - DMPG) show that these lipids enhance the rate of coagulation by shortening the reaction of coagulation, and stimulating platelet and fibrinogen activation. Negatively charged lipids are also observed to shorten the clotting time by up to 25% and cause 45 - 85% platelet and fibrinogen activation.   These results are in agreement with other previous studies showing that negatively charged lipids

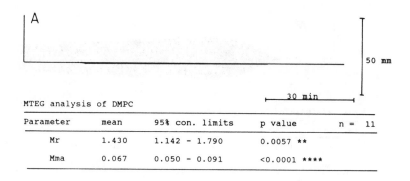

MTEG analysis of DMPC

Parameter	mean	95% con. limits	p value	n = 11
Mr	1.430	1.142 - 1.790	0.0057 **	
Mma	0.067	0.050 - 0.091	<0.0001 ****	

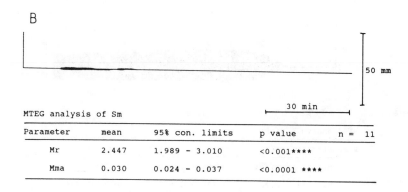

MTEG analysis of Sm

Parameter	mean	95% con. limits	p value	n = 11
Mr	2.447	1.989 - 3.010	<0.001****	
Mma	0.030	0.024 - 0.037	<0.0001 ****	

Fig. 3: MTEG analysis of phosphorylcholine containing phospholipids. Typical traces are shown for (A) dimyristoylphosphatidylcholine - DMPC and (B) sphingomyelin - Sm.

have the ability to enhance prothrombin activation and Factor X activation (Froman and Nemerson, 1986). DPPA and the RBC_i lipids particularly bind factors XII, VII and IX.

Positively charged surfaces in the MTEG test, with lipids such as stearylamine (St) and hexadecylamine (HDA) prolong the whole blood clotting time (increased by 39 - 43%) and cause up to 45% platelet and fibrinogen activation (Fig. 2). Surface charge has been shown to affect platelet aggregation which is enhanced in the presence of cationic compared to anionic polymers (Kataoka *et al.*, 1978).

Increasing the concentrations of amines incorporated into DPPC liposomes prolongs the Stypven time of citrated plasma suggesting that positively charged surfaces can inhibit coagulation. It is possible that clotting factors bind indirectly to the positively charged surfaces but then have limited ability to form the protein aggregates necessary to generate the 'tenase' and 'prothrombinase' complexes. Furthermore, the Clotting Factor Assay show that positively charged lipids do bind to many proteins.

The MTEG studies with lipid PC surfaces such as dimyristoylphosphatidylcholine (DMPC) and sphingomyelin (Sm) show little or no blood coagulation (Fig. 3). The clotting times of these surfaces are extended by up to three times that of the controls whilst platelet and fibrinogen activation is less than 6%. Furthermore PC-containing liposomes have no effect on the Stypven time (Zwaal et al., 1977; Hayward and Chapman 1984) and no significant binding of clotting factors to PC-containing liposomes was observed. Recent studies by Chonn et. al. (Chonn et al., 1990) also show that phosphorylcholine liposomes do not affect the complement system, whereas negatively and positively charged phospholipids activate the classical and alternative pathway respectively. The zwitterion PC polar group appears to act as a completely neutral surface with regard to blood coagulation and blood factors.

A question which we have been considering is what are the properties of the PC group which gives rise to the good haemocompatible characteristics.

Biophysical studies of the PC group using NMR show that the phosphorylcholine group only weakly interacts with calcium ions and does not interact with monovalent ions up to 1M NaCl (Altenbach and Seelig, 1984). This implies that the PC group will not readily support the formation of calcium bridges with blood clotting proteins. Altenbach and Seelig (1984) also demonstrated that the PC polar head groups are oriented parallel to the lipid membrane surface, and may therefore be less accessible for interaction with blood proteins. The high hydration level of the PC polar head group, some 20 moles per lipid head group (Chapman et al., 1967), may also help to limit the interactions between the blood proteins and the lipid membrane.

New polymers: In our laboratory we have been synthesising new methactylate polymers containing the PC group and studying cell interaction - with these polymers..

Acknowledgements: This work was supported by research grants from the Wellcome Trust and the SERC, Medical Engineering Sub Committee.

REFERENCES

Altenbach, C., and Seelig, J. (1984) Biochemistry, 23, 3913-3920.

Alvarez, J., Haris, P.I., Lee, D.C. and Chapman, D. (1987) Biochimica et Biophysica Acta, 916, 5-12.

Bird, R. le R., Hall, B., Chapman, D. and Hobbs, K.E.F. (1988) Tromb. Res. 51, 471-483.

Chapman, D. and Benga, G. (1984) In: Biological Membranes, vol.5 (D. Chapman, Ed), Academic Press, London, pp. 1-56.

Chapman, D., Williams, R.M., and Ladbrooke, B.D. (1967) Chem. Phys. Lipids, 1, 445-475.

Chonn, A., Lafleur, M., Kitson, N., Deriac, D.V., Cullis, P.R. (1990) 10th International Biophysics Congress, Vancouver, Canada.

Froman, S.D. and Nemerson, Y. (1986) Proc. Natl. Acad. Sci., 83, 4675-4679.

Hall, B., Hall, C.J., Hutton, R.A., Bird, R.le R. and Chapman, D. (In press).

Haris, P.I., Chapman, D., Harrison, R.A., Smith, K.F. and Perkins, S.J. (1990) Biochemistry, 29, 1377-1380.

Haris, P.I., Lee, D.C. and Chapman, D. (1986) Biochim. Biophys. Acta., 874, 255-265.

Haris, P.I. and Chapman, D. (1989a) Biochemical Society Transactions, 17, 161-162.

Haris, P.I., Coke, M. and Chapman, D. (1989b) Biochimica et Biophysica Acta, 995, 160-167.

Hayward, J.A. and Chapman, D. (1984) Biomaterials, 5, 135-142.

Johnston, D.S., Sanghera, S., Manjou-Rubio, A. and Chapman, D. (1980) Biochim. Biophys. Acta, 602, 213-216.

Kataoka, K., Akaike, T., Sakurai, Y., and Tsuruta, T. (1978) Makromol. Chem., 179, 1121-1124.

Kennedy, D.F., Crisma, M., Toniolo, C. and Chapman, D. (1991) Biochemistry, 30 6541-6548

Kennedy, D.F., Slotboom, A.J., Haas, G.H. de and Chapman, D. (1990) Biochemica Biophysica Acta, 1040, 317-326.

Lee, D.C., Hayward, J.A., Restall, C.J. and Chapman, D. (1985) Biochemistry, 24, 4364-4373.

Lee, D.C., Haris, P.I., Chapman, D. and Mitchell, R.C. (1989) In: 'Spectroscopy of Biological Molecules - State of the Art' (A. Bertoluzza, C. Fagnano and P. Monti eds) Societa Editrice Esculapio, Bologna, 57-58.

Lee, D.C., Haris, P.I., Chapman, D. and Mitchell, R.C. (1990) Biochemistry, 29, No. 39, 9185-9193.

Lee, D.C., Haris, P.I., Chapman, D. and Mitchell, R.C. (1991) 4th European Conference Spectroscopy of Biological Molecules, York.

Levitt, M. and Greer, J. (1977) J. Mol. Biol., 114, 181-293.

Mitchell, R.C., Haris, P.I., Fallowfield, C., Keeling, D.J. and Chapman, D. (1988) Biochimica et Biophysica Acta, 941, 31-38.

Perkins, S.J., Nealis, A.S., Haris, P.I., Chapman, D., Goundis, D. and Reid K.B.M. (1989) Biochemistry, 28, 7176-7182.

Perkins, S.J., Haris, P.I., Sim, R.B. and Chapman, D. (1988) Biochemistry, 27, 4004-4012, American Chemical Society.

Sarver, R.W. and Krueger, W.C. (1991) Analytical Biochemistry, 194, 89-100.

Villalain, J, Gomez-Fernandez, J.C., Jackson, M. and Chapman, D. (1989) Biochimica et Biophysica Acta, 978, 305-312.

Yool, A.J. and Schwarz, T.L. (1991) Nature, 349, 700-704.

Zwaal, R.F.A., Comfurius, P., and van Deenen, L.L.M. (1977) Nature (London), 268, 360-362.

Progress in Membrane Biotechnology
Gomez-Fernandez/Chapman/Packer (eds.)
© 1991 Birkhäuser Verlag Basel/Switzerland

DYNAMIC X-RAY DIFFRACTION STUDIES OF PHASE TRANSITIONS IN LIPID-WATER SYSTEMS

P.J. Quinn and L.J. Lis

Section of Biochemistry, Division of Biomolecular Sciences, King's College London, Campden Hill, London W8 7AH, UK and Section of Hematology/Oncology, The Chicago Medical School, VA Medical Centre, North Chicago, IL 60064, USA

SUMMARY: The application of time-resolved X-ray diffraction methods to characterise phase transitions in lipid-water systems is described.   Used in conjunction with thermal and densitometric measurements it is possible to establish the precise mechanisms of phase transitions and the thermodynamic order of these processes. The construction of dynamic phase diagrams which define the parameters for creation of metastable phases can also be accomplished using these methods.   The time resolution of the structural changes that can be observed using synchrotron X-radiation is well within the relaxation time of most phase transitions which take place in the order of a few seconds. Examples of the application of these methods are used to illustrate the utility of this combined approach to characterise phase transitions in these model membrane systems.

One of the characteristic features of the lipid components of biological membranes is the remarkable diversity of molecular species present. This varibility must be considered against the background of the primary function of the lipids which is generally considered to be to form a lipid bilayer which acts as a matrix for the support and integration of the different membrane proteins.

Our knowledge of the purpose of the complex lipid composition is rudimentary although it has been appreciated for some time that reconstitution of activity of many membrane proteins has a minimal requirement for lipids that form a fluid bilayer arrangement. This is fulfilled by the general amphipathic character of the lipids in which the hydrophobic domain of the molecule is separated discretely from the polar group and the presence of <u>cis</u> unsaturated double bonds or bulky branch groups in the substituent acyl chains. The presence of unsaturated fatty acyl residues on some classes of membrane lipid, notably, monohexosyldiacylglycerols and phosphatidylethanolamines transforms these lipids from bilayer-forming to non-bilayer forming lipids. This means that when such lipids are dispersed in aqueous systems they typically form an hexagonal-II phase at physiological temperatures. Admixture with bilayer forming lipids often leads to phase separations in much the same way that bilayer domains of gel and fluid lipids can be created when binary mixtures of lipids with large differences in gel to liquid-crystalline phase transition temperatures are cooled.

Studies of the phase behaviour of membrane lipids and their synthetic counterparts is fundamental to our understanding of the role that these important components play in membrane structure and stability. It is apparent from the many studies that have been reported that membrane lipids, when dispersed in aqueous systems, display a rich polymorphism as a function of both temperature and hydration. Moreover, many processes are metastable and are transform with time varying from seconds to weeks or more to relativly stable phases.

Polymorphic behaviour of these materials is usually characterised by calorimetry which can provide information relevant to the temperature and enthalpy associated with transitions between phases. The phases themselves can be assigned by determination of structural parameters using X-ray diffraction, nuclear magnetic resonance spectroscopy, vibrational spectroscopy etc. Such methods are essentially static and the sample must be at or near to equilibrium conditions. By contrast, scanning calorimetry is a dynamic method in which metastable phases are revealed.

In order to reconcile the thermal and structural parameters associated with polymorphic phase behaviour X-ray diffraction studies have been performed under identical conditions used to provide thermal data. The high X-ray flux required to record diffraction patterns on the time-scale of ms has been achieved using synchrotron radiation sources and studies from a number of these facilities has been reported (see reviews by Laggner, 1986, 1988; Caffrey, 1989). Studies of the mechanism and kinetics of phase transitions in lipid-water systems using a combination of thermal and time-resolved X-ray diffraction methods have also been performed (Tenchov & Quinn, 1989; Lis & Quinn, 1991).

METHODS

CALORIMETRY: A number of methods are available to characterise the transition between phases formed by polar lipids dispersed in aqueous systems. The most common method is to measure enthalpy changes by some form of calorimetry. Differential scanning calorimetry (Silvius, 1982; McElhaney, 1982; Small, 1986) is one method which accurately measures the heat required to sustain a constant change in temperature within a lipid dispersion compared with a reference sample of similar heat capacity.

Although differential scanning calorimetry provides an accurate temperature at which the conversion from one phase to another takes place, as well as a measure of the enthalphy of the particular transition, it does not provide any structural information about the transition. However, useful information on phase mixing and compound formation can be derived from the method. Moreover, the relative co-operativity of a phase transition can be evaluated directly from the thermogram since it is related to the sharpness of the excess specific heat absorption curve of an endothermic transition. This is generally expressed as the temperature width at half height of the heat absorption peak or as the temperature difference between the onset or low temperature boundary of the phase transition and the completion of the upper limit of the transition temperature. Using this criterion very pure single component lipids exhibit highly co-operative chain melting endotherms with temperature widths at half height of less than $0.1°$ but this may extend to 10-15° for complex mixtures and biological membranes.

DYNAMIC X_RAY DIFFRACTION: Conventional X-ray diffraction methods have been employed for over 40 years to characterise the structure of lipid-water systems of relevance to biological membranes (Luzzati, 1968; Shipley, 1973; Blaurock, 1983). Because the flux of X-rays produced, even by modern rotating anode generators, is relatively low accumulation of scattered X-rays over minutes or even hours is required to produce recognisable diffraction patterns. With the advent of synchrotron radiation sources capable of producing light of extremely high brilliance it has become possible to obtain structural information on lipid-water systems in real time during phase transition events.

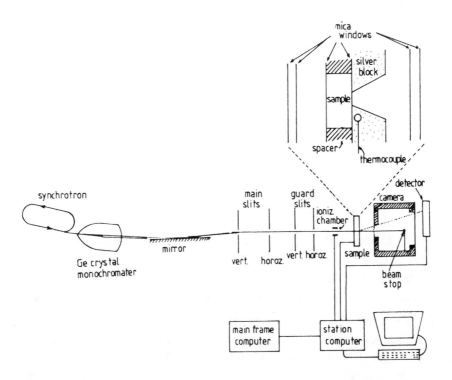

Fig. 1. Schematic diagram of the camera configuration used on station 8.2 of the Daresbury synchrotron radiation source showing detail of the sample arrangement.

Compared with conventional X-ray sources the time resolution can be improved by some three or four orders of magnitude in recording diffraction data. This means that structural changes associated with thermal phase transitions can be monitored on a time scale of milliseconds. A time resolution of this order is sufficient to determine the absolute rate of relaxation processes that are associated with phase transitions driven by temperature or pressure jumps (Caffrey, 1989). Furthermore, greater resolution of structural transitions during slow passage through phase transitions can be achieved (Tenchov, Yao & Hatta, 1989).

The experiments described in this chapter were performed on station 8.2 of the Daresbury Synchrotron Radiation Source. A diagram of the camera configuration used to perform the real-time measurements is shown in Fig. 1. The Ge crystal monochromator provides X-rays of wavelength of 0.15nm and these are horizontally focused and columnated by the mirror and slits respectively. X-ray flux was monitored by an ionization chamber and the sample temperature controlled by a modified cryostage to give temperature gradients between $0.2°$ and $100°$/min. over a temperature span of 125K to >600K. The scattered X-rays from unoriented samples were recorded using a linear delay line detector; up to 255 sequential diffraction patterns could be acquired over appropriate times with a dead time of $20\mu s$ between frames.

APPLICATIONS

METASTABLE PHASE CHARACTERISATION: Many examples of metastable phase formation in lipid-water systems have been described. Perhaps the most common type of metastability is the lamellar gel phase which often converts to a stable lamellar crystal or subgel phase in which the hydrocarbon chains are packed on a more regular lattice and water may be expelled from the mesophase. The relaxation process depends on the temperature, water concentration and time of storage.

A metastable gel phase that cannot easily be characterised by differential scanning calorimetry is illustrated by the heating

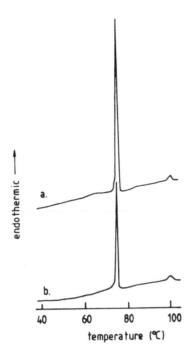

Fig. 2. Differential scanning calorimetric heating scans of an aqueous dispersion of distearoylphosphatidylethanolamine equilibrated for 3 days at 20°C (a) and recorded immediatly after the first scan (b). Data from Williams et al (1991).

thermograms of distearoylphosphatidylethanolamine dispersed in excess water (Fig. 2) and is typical of saturated phosphatidylethanolamines (Seddon et al, 1983; Chang and Epand, 1983; Mantsch et al, 1983). When the distearoyl derivative is hydrated at 60°C and subsequently heated (Fig. 1, thermogram a) there is a conspicuous endotherm at 73°C with an enthalpy of 16.8 kJ. mole^{-1} followed by a much smaller endotherm at about 100°C with an enthalpy of only 0.62 kJ.mole^{-1}. If the sample is cooled and immediately reheated the first endotherm appears at almost the same temperature (72°C) but the enthalpy is half the original value (8.6 kJ.mole^{-1}). The smaller, high-temperature endotherm remains unchanged although there is a slight broadening of the transition indicating some loss in cooperativity.

18

To characterise the phases formed by the phospholipid, time-resolved X-ray diffraction measurements were performed and diffraction patterns recorded during the main endothermic transition are shown in Fig. 3. The phase that forms on initial hydration is a lamellar crystal or subgel phase which has a lamellar repeat spacing of 5.25nm. The chain packing, indexed by the wide-angle scattering, shows a prominent band at a spacing corresponding to 0.437nm and two less prominent peaks centred at 0.393 and 0.372nm respectively. The comparitively small lamellar repeat of this phase is consistent with a tilting of the hydrocarbon chains with respect to the bilayer normal in a way

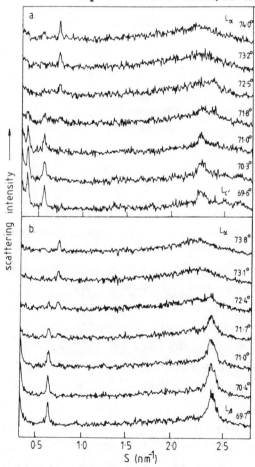

Fig. 3. X-ray scattering intensity vs reciprocal spacing recorded over the transition region in samples identical to those shown in Fig. 2. Data from Williams et al (1991)

typical of that of the B_1 form of dilauorylphosphatidylethanolamine reported by Seddon et al (1983) but with a somewhat different packing arrangement of the hydrocarbon chains as judged by the precise positions of the wide-angle diffraction maxima. Heating the phospholipid in the subgel phase results in a reduction in intensity of the wide-angle reflections and their replacement at the transition temperature by a broad diffuse scattering band centred at 0.44nm typical of disordered hydrocarbon chains in a liquid-crystalline configuration (Fig. 3a). The regular smectic subgel phase is characterised by the sharp second and third-order lamellar repeat bands which decrease in intensity in tandem with the wide-angle diffraction peaks during the transition. They are replaced by the lamellar liquid-crystal phase characterised by a sharp fourth-order peak indexing a spacing of 5.2nm. The transition appears to be two-state with coexistance of the subgel and lamellar liquid-crystal phases during the transition with no evidence of intermediate structures. The changes observed in the X-ray pattern upon heating through the high-temperature endotherm involve only the low-angle reflections. The first order reflection shows an increase in spacing from 5.2nm to 6.1nm and the higher-order spacings change from $d:d/2:d/3:d/4$ to $d:d/\sqrt{3}:d/\sqrt{4}:d/\sqrt{7}$ characteristic of a non-lamellar phase. Although higher orders of reflection are required to provide a more confident assignment of phase, freeze-fracture electron microscopy indicated some type of hexagonal phase and hexagonal-II structure with the polar groups oriented about tubes of water is the most likely arrangement.

The x-ray pattern which forms on cooling the hexagonal phase (Fig.3b) shows a lamellar arrangement with a repeat spacing of 6.25nm and a single sharp reflection centred at 0.41nm in the wide-angle region indicative of a lamellar gel phase. This phase was found to be stable over four days at $20°C$. Restoration of the subgel phase following high-temperature treatment takes place during subsequent low-temperature storage at a rate dependent on the chain length of the phosphatidylethanolamine. Thus the subgel phase is formed during 10 days for the dilauoryl derivative and increases to 20 days for the dimyristoyl derivative; presumably longer periods of storage would be required for the dipalmitoyl and distearoyl forms. The X-ray patterns recorded during heating the lamellar gel phase confirm that the corresponding endotherm is due to a lamellar gel to lamellar liquid-crystalline phase transition. The patterns recorded during the high-temperature transition were identical to those observed during the first heating, and indicate a lamellar liquid-crystalline to hexagonal-II phase transition.

Relaxation from lamellar gel to subgel phases occurs on a faster time scale in the case of distearoylmonogalactosyldiacylglycerol (Lis & Quinn, 1986). This transition can be seen in the X-ray patterns recorded during isothermal storage at 20°C of a hydrated sample that had been cooled from the liquid-crystal phase (Fig.4). The initial phase is a lamellar-gel with a repeat spacing of 6.25nm and a single wide-angle diffraction peak centred at a spacing of 0.41nm. Two broad bands are discernable in an intermediate position of the pattern with spacings of 0.61 and 0.69nm respectively which arise from a rectangular packing of the galactose residues. Sequential patterns of 6s duration are shown

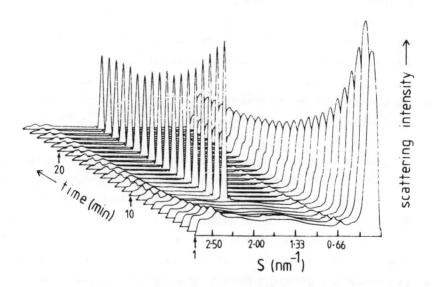

Fig. 4. X-ray diffraction patterns recorded isothermally at 20°C following cooling of a dispersion of distearoylmonogalactosyl-diacylglycerol from 60°C. Data from Lis and Quinn, (1986).

(one in ten frames of the complete data set) during incubation. After 8 minutes there is a dramatic change in the diffraction pattern which takes place in a cooperative manner over a period of about 30s. The transformation involves the replacement of the single wide-angle reflection by two peaks corresponding to spacings of 0.40nm and 0.38nm respectively and the simultaneous transition of the bands at intermediate spacing to a single sharp diffraction maximum centred at 0.60nm. These changes are interpreted as a reorientation of the molecules within the bilayer so that they pack into a phase of more ordered structure. This is consistent with the enthalphy change associated with the transition which is associated with an exothermic reaction. The creation of a more ordered acyl chain subcell in the subgel phase suggests that the ordering of the chains is the primary factor responsible for driving the transition. The concomitant disordering of the relatively bulky sugar residues must be required to reduce steric hinderance to accommodate the acyl chain packing into a rectangular lattice. It may also be seen in Fig.4 that there is a change in the mesophase spacing during storage. The subgel phase formed initially has a lamellar repeat of 7.41nm and this transforms eventually into another subgel phase with a lamellar repeat of about 6nm.

TRANSITION PATHWAY: The structural rearrangements that take place during transition between equilibrium phases can vary from simple two-state transitions to highly complex transitions involving metastable intermediate phases. The latter types of transition are often associated with enthalpic events that are difficult to interpret. Time-resolved X-ray diffraction methods can be used in such systems to characterise the precise transition pathway and to identify intermediate phases that occur during the transition.

An example of a transition exhibiting complex thermal behaviour that has been examined by dynamic X-ray methods is the hexagonal-II to lamellar transition in the distearoylphosphatidylethanolamine/ glycerol system. In dispersions of the phospholipid in excess glycerol (>70wt% glycerol) heating thermograms exhibit complex endothermic and exothermic behaviour (Williams et al, 1991) indicating that subgel phases may be present as intermediates in the overall transition pathway.   Examination by dynamic X-ray diffraction of a dispersion of distearoylphosphatidylethanolamine in 90wt% glycerol has confirmed the existence of a subgel phase between lamellar gel and hexagonal-II phases.   The transition sequence is illustrated in Fig.5 which shows diffraction patterns recorded in a dispersion cooled to 60°C and immediately reheated at 5°/min through the transition region.   The presence of the subgel phase is identified by the multiple wide-angle scattering peaks that are due to the non-hexagonal packing of the hydrocarbon chains and also the lamellar repeat spacing is different compared to that of the lamallar gel phase.

More   complicated   mechanisms   of   relaxation   of   the disteroylphosphatidylethanolamine/ glycerol system are observed upon cooling the hexagonal-II equilibrium high-temperature phase.

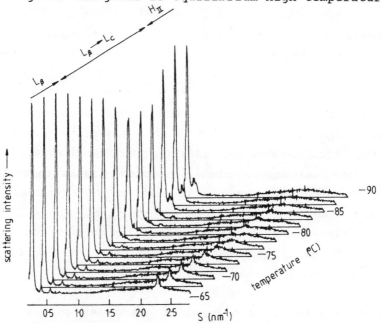

Fig. 5.   A sequence of X-ray scattering patterns recorded in a dispersion of distearoylphosphatidylethanolamine in excess 90wt% glycerol cooled to 60°C and reheated again at 5°/min.   The phase assignments are indicated.   Data from Williams et al (1991).

Examination of dispersions cooled 20°C and stored over several days
has indicated a transition from a lamellar gel phase to one of two
possible subgel phases one of which has hydrocarbon chains oriented
normal to the plane of the bilayer and the other with tilted
chains.   The factors governing which of the subgel phases that
will form are presently obscure.

Dispersions of the phospholipid in pure glycerol cooled from the
hexagonal-II phase have shown evidencefor coexistance of the
hexagonal-II phase with the subgel phase during the onset of the
transition phase but as the apparent transition $H_{II} \rightarrow L_C$ proceeds, a
point is reached where the remaining lipid in the H_{II} phase
undergoes a 2-state $H_{II} \rightarrow L_\beta$ phase transition.   These events are shown
clearly in the data presented in Fig.6 in which a sample of
distearoylphosphatidylethanolamine dispersed in excess glycerol was
cooled from 90°C at a rate of 5°/min.   The appearance of second and
third-order lamellar repeat spacings of the subgel phase are first
observed in diffraction patterns recorded at 85°C and are very
conspicuous at 68.2°C.   The remaining phospholipid in hexagonal-II

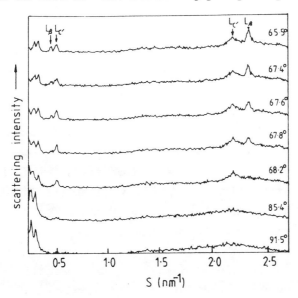

Fig. 6.   Selected diffraction patterns recorded during cooling of
a mixture of distearoylphosphatidylethanolamine in excess glycerol
from 90°C at a rate of 5°/min.   Data from Williams et al   (1991).

phase, characterised by the $d/\sqrt{3}$ second-order reflection of the low-angle repeat, is converted to spacing of 0.41nm at a temperature of 68°C. A similar coexistance of the subgel and hexagonal-II phase in partially hydrated specimens of diarachidonylphosphatidylethanolamine has also been reported (Seddon et al, 1984).

PHASE TRANSITION MECHANISMS: Structural changes associated with a transition between phases can be observed both in terms of mesophase structure and order and packing of the hydrocarbon chains. Calorimetry measurements performed under identical conditions enable correlation between specific heat and structural changes. Such is the versatility of the approach that structural changes have been observed that involve no detectable changes in specific heat capacity, even using high-sensitivity differential scanning calormetry (Tamura-Lis, et al, 1990).

Application of classical thermodynamic principles has allowed a prediction of the structural changes that take place according to the order of the phase transition (Callen, 1961). Accordingly, first-order phase transitions proceed with coexistance of the initial and final states while higher-order transitions are characterized by the presence of one or more independent transitory states. Application of the X-ray method to detect phase states, which are essentially bulk properties of the system, obviously depend on the ability to characterize a particular phase state from the scattering pattern. The factors governing the limits of detection of a structural state depend on the domain size and the order within the diffracting lattice as well as the lifetime of the phase state. Tenchov et al (1989) have introduced a scaler concept to provide an operational definition of phase transition mechanisms. Thus a coexistance of phase states may be "large scale" if the three dimensional domain sizes are sufficiently large to produce a superposition of two discrete phases in the diffraction pattern. An example of such a transition was found to be the interdigitated lamellar gel to liquid-crystalline phase transition of dipalmitoylphosphatidylcholine dispersed in ethanol-water solution. In this system there was a range of temperatures within the transition region where the lamellar repeat reflections were found to coexist. This was in contrast to the main lamellar gel to liquid-crystalline phase transition of the phospholipid dispersed in water where under equilibrium conditions the transition region was characterised by a broadening of the lamellar reflections due to disorder in the bilayer stacking and partial loss of correlation in lamellar structure as the hydrocarbon chains become disordered. This situation is not the case under non-equilibrium conditions, such as experienced in temperature jump experiments where phase coexistence can be clearly observed in this phospholipid (Laggner et al, 1987).

The studies performed by Tenchov et al (1989) and Laggner et al (1987) were restricted to phase characterization dependent on low-angle scattering. It should be noted that where both low-angle spacings and chain order are recorded simultaneously transition order can be seen to involve structural changes on two levels. Phase transitions that do not involve changes in mesophase d-spacing (commensurate structures) but only a change in order or packing of the hydrocarbon chains have been reported and include dipalmitoylphosphatidylcholine dispersed in ethanol-water solution (Tenchov et al, 1989). It is assumed that a decrease in bilayer thickness associated with disordering of the hydrocarbon chains is precisely compensated by an increase in the water layer thickness. Phase transitions where d-spacings change during the transition (incommensurate structures) combine structural rearrangements at the level of the mesophase structure as well as chain configuration. In both cases, where mesophase structure is associated with a change in hydration, water removal or penetration into the lattice may represent a rate limiting step producing an apparent higher-order process.

An example of a higher-order transition characterized by a change in packing of the hydrocarbon chains is the subgel transition in dipalmitoylphosphatidylcholine (Tenchov et al, 1987). This is illustrated in Fig.7 which shows wide-angle scattering intensity

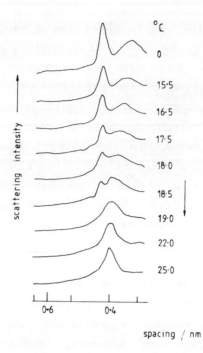

spacing / nm

Fig. 7. X-ray scattering intensity recorded at wide angles during heating a dispersion of dipalmitoylphosphatidylcholine in the subgel phase through the subtransition at 5°/min. Data from Tenchov et al (1987).

recorded during heating of an aqueous dispersion of dipalmitoylphosphatidylcholine equilibrated at 4°C through the subgel transition centred about 18°C. It can be seen that during the subtransition two prominent wide-angle reflections, characteristic of the low-temperature subgel phase, gradually change such that a sharp peak at a spacing of 0.43nm decreases in intensity until it finally disappears while a broader peak located initially at a spacing of 0.375nm progressively shifts to an eventual spacing of 0.41nm. These changes have been interpreted as a lateral deformation of the acyl-chain packing subcell as the chains begin to rotate in transition from the crystal to the gel state where the chains pack on a regular hexagonal array. An increase in the lamellar repeat distance from 6.0 to 6.4nm accompanies the changes in hydrocarbon chain packing and, in common with two-state lamellar transitions, changes in packing order of the chains dictate the change in mesophase structure and water movements through the phase may be a rate limiting process in the subtransition.

KINETICS OF PHASE TRANSITIONS: Because of the intensity of synchrotron X-rays it is possible to record diffraction patterns in the order of a few ms. With a time resolution in this domain it is possible to observe relaxation processes associated with the transition of lipid-water systems from one phase to another. The data obtained using stopped flow, temperature or pressure jump methods of inducing phase transitions can be used as a basis for formulating, testing and refining phase transition mechanisms. Moreover, kinetic measurements reveal details of molecular structure and phase composition that modulate the transition mechanism and rate. The application of synchrotron radiation to examine the kinetics of phase transitions in lipid-water systems has been reviewed by Caffrey (1989).

Measurements of phase transition kinetics of a wide variety of lipid-water systems using different methods to induce the transitions have indicated that most transitions occur on a time scale within about 1-5s. Transitions of metastable phases into more stable phases take place over much longer times as referred to above. One of the problems associated with temperature jump methods is to prevent temperature gradients developing within the sample. Various methods have been used to create temperature jumps including circulating water baths (Caffrey, 1984), air stream (Caffrey and Bilderback, 1984) microwave irradiation (Caffrey et al, 1990), pulsed laser light irradiation and absorption (Kreichbaum et al, 1989; Laggner et al, 1989) or Peltier devices (Ranck et al, 1984) each with their own particular advantages. Pressure jump methods have not yet been employed extensively mainly because of the X-ray opacity of materials required to sustain the high pressures required to induce phase transitions; the method, however, has the distinct advantage of avoiding the development of gradients through the sample.

CONCLUSION: The utility of employing a combination of thermal and structural measurements in characterising phase transitions in model membrane systems is clearly established. Time-resolved X-ray diffraction exploiting the high brightness of synchrotron light sources used in conjunction with differential scanning calorimetry and scanning densitometry enable a precise determination of structural changes associated with measured thermodynamic events. The determination of the mechanism of transitions and hence the thermodynamic order can be made, particularly if low temperature scan rates are employed in recording the x-ray scattering data. Furthermore, the transition mechanism can be used to provide information on the existence of thermodynamically-predicted intermediate phases and the effect of domains on the transition process. Finally, construction of dynamic phase diagrams using these time-resolved methods can characterize the extent of metastability of these lipid-water systems.

Acknowledgements: The authors thank Wim Bras for assistance with the synchrotron X-ray diffraction measurements and the Science and Engineering Research Council (U.K.) for financial support.

REFERENCES
Blaurock, A.E. (1983) Biochim. Biophys. Acta, 650, 167-184.
Caffrey, M. (1984) Nuc. Instrum. Meth., 222, 329-338.
Caffrey, M. (1989) Ann. Rev. Biophys. Biophys. Chem., 18, 159-186.
Caffrey, M. and Bilderback, D.H. (1984) Biophys. J., 45, 627-631.
Caffrey, M., Magen, R.L., Hummel, B. and Zhang, J. (1990) Biophys. J., 58, 21-29.
Callen, H.B. (1961) Thermodynamics, Wiley, New York.
Chang, H. and Epand, R.M. (1983) Biochim. Biophys. Acta, 728, 319-324.
Kreichbaum, M., Rapp, G., Hendrix, J. and Laggner, P. (1989) Rev. Sci. Instrum., 60, 2541-2544.
Laggner, P. (1986) In: Structural Biological Uses of X-ray Absorption, Scattering and Diffraction (H. Bartunik and B. Chance, Eds.) Academic Press, London, pp 171-182.
Laggner, P. (1988) Top. Current Chem., 145, 173-202.
Laggner, P., Kreichbaum, M., Hermetter, A., Paltauf, F., Hendrix, J. and Rapp, G. (1989) Prog. Colloid Polymer. Sci., 79, 33-37.
Laggner, P., Lohner, K. and Müller, K. (1987) Mol. Cryst. Liq. Cryst., 151, 373-388.

Lis, L.J. and Quinn, P.J. (1986) Biochim. Biophys. Acta, <u>862</u>, 81-86.

Lis, L. J. and Quinn, P. J. (1991) Acta Cryst., <u>24</u>, 48-60.

Luzzati, V. (1968) In: Biological Membranes, Vol. 1 (D. Chapman, Ed.), Academic Press, London, pp. 71-123.

McElhaney, R.N. (1986) Biochim. Biophys. Acta, <u>864</u>, 361-421.

Mantsch, H.H., Hsi, S.C., Butler, K.W. and Cameron, D.G. (1983) Biochim. Biophys. Acta, <u>728</u>, 325-330.

Ranck, J.L., Lattelier, L., Schechter, E., Krop, B., Pernot, P. and Tardieu, A. (1984) Biochemistry, <u>23</u>, 4955-4961.

Seddon, J.M., Harlos, K. and Marsh, D. (1983) J. Biol. Chem., <u>258</u>, 3850-3854.

Seddon, J.M., Cevc, G., Kaye, R.D. and Marsh, D. (1984) Biochemistry, <u>23</u>, 2634-2644.

Shipley, G.G. (1973) In: Biological Membranes, Vol. 2 (D. Chapman and D.F.H. Wallach, Eds.), Academic Press, London, pp. 1-89.

Silvius, J.R. (1982) In: Lipid-Protein Interactions, Vol. 2 (P.C. Jost and H.O. Griffith, Eds.) Wiley, New York, pp. 239-281.

Small, D.M. (1986) In: Handbook of Lipid Research, Vol. 4 (D.J. Hanahan, Ed.), pp. 1-672.

Tamura-Lis, W, Lis, L.J., Qadry, S. and Quinn, P.J. (1990) Mol. Cryst. Liq. Cryst., <u>178</u>, 79-88.

Tenchov, B.G., Lis, L.J. and Quinn, P.J. (1987) Biochim. Biophys. Acta, <u>897</u>, 143-151.

Tenchov, B. G. and Quinn, P. J. (1989) Liq. Cryst., <u>6</u>, 1691-1695

Tenchov, B.G., Yao, H. and Hatta, I. (1989) Biophys. J. <u>56</u>, 757-768.

Williams, W.P., Quinn, P.J., Tsonev, L.I. and Koynova, R.D. (1991) Biochim. Biophys. Acta, <u>1062</u>, 123-132.

Progress in Membrane Biotechnology
Gomez-Fernandez/Chapman/Packer (eds.)
© 1991 Birkhäuser Verlag Basel/Switzerland

Phase Behavior of Phosphatidic Acid

C.-C. Yin, B.-Z. Lin and H. Hauser

Laboratorium für Biochemie, ETH Zürich, ETH-Zentrum, CH-8092 Zürich, Switzerland

SUMMARY: Phosphatidic acids have a propensity for forming smectic (lamellar) phases. Under physiological conditions phosphatidic acids bear one negative charge/molecule and phosphatidate bilayers swell continuously with increasing water content similar to other negatively charged lipid bilayers. If smectic (lamellar) phases of egg phosphatidic acid or dilauroylphosphatidic acid in excess water are energized, e.g. by ultrasonication or by pH-jump treatment, small unilamellar vesicles form. The thermodynamic stability of these vesicles was studied by making use of the carboxyfluorescein assay. It was found that small unilamellar phosphatidate vesicles of a diameter of 20 to 60 nm are stabilized by a proton gradient of 2 - 4 pH units (pH outside alkaline, pH inside neutral or acid). These vesicles are, however, thermodynamically unstable in the absence of a pH gradient.

Introduction: Phosphatidic acids are important intermediates in the metabolism of triacylglycerols and phospholipids (Hjelmstad & Bell, 1991). The phosphatidic acid content of tissue, plasma and subcellular membranes appears to be carefully regulated and usually does not exceed a few percent. Phosphatidic acids bear one negative charge at neutral pH and have a tendency to form smectic (lamellar) phases. Mixtures of phosphatidic acids with other phospholipids, particularly phosphatidylcholine, have been the subject of extensive studies. Information concerning the phase behavior and properties of pure phosphatidic acids differing in hydrocarbon chain length and degree of unsaturation is still scarce.

MATERIALS AND METHODS

Egg phosphatidic acid was purchased from Lipid Products (Surrey, U.K.). The disodium salt of 1,2-dilauroyl-sn-glycero-3-phosphoric acid (DLPA) was synthesized by Mr. R. Berchtold (Biochemisches Labor, Bern, Switzerland). The phospholipids used in this study were pure by thin-layer chromatography standard.

The preparation of sonicated phospholipid dispersions and the pH-jump method were described before (Hauser & Gains,1982; Hauser et al.,1990). Small unilamellar vesicles of phosphatidic acid of diameter of about 20 nm were prepared by either sonication or pH-jump. If carboxyfluorescein was to be incorporated into the vesicle cavity, the chromophore was added at self-quenching concentration of ~ 50 mM to the dispersion medium (10 mM Tris buffer pH 7.2 containing 0.15 M NaCl, 0.02% NaN_3). External carboxyfluorescein was separated from carboxyfluorescein-loaded vesicles by gel filtration on a Sephadex G-50 column (35 x 1.2 cm) equilibrated and run with 10 mM Tris buffer pH 7.2 containing 0.1 M NaCl, 0.02% NaN_3. Phosphatidic acid vesicles were labeled with a trace of 1,2-dipalmitoyl-sn-phosphatidyl [N-methyl-^{3}H]choline (^{3}H-DPPC) obtained from Amersham, U.K. The phospholipid concentration in the effluent was determined by counting radioactivities.

Gel filtration on calibrated Sepharose CL-4B (50 x 0.9 cm) was carried out as decribed (Schurtenbenger & Hauser, 1984). The methods of differential scanning calorimetry, X-ray diffraction and ^{31}P NMR were also described before (Casal et al., 1990).

Fluorescence Spectroscopy: The following fluorescence assay was used to monitor the thermodynamic stability of small unilamellar phosphatidic acid vesicles. The assay is based on carboxyfluorescein encapsulated in the vesicle cavity as described above. Furthermore, the assay is based on the assumption that the thermodynamically unstable vesicles fuse and that fusion is accompanied by leakage of the entrapped carboxyfluorescein. Leakage of the chromophore into the bulk phase is readily detected by an increase in fluorescence intensity. Fluorescence intensities were measured with an Aminco SPF-500 fluorimeter (excitation at 470 nm, bandpass = 5 nm, emission at 520 nm, bandpass = 10 nm). The 100% value of fluorescence intensity was obtained when all entrapped carboxyfluorescein was released, e.g., after solubilizing the small unilamellar phospholipid vesicles with 2% sodium cholate.

RESULTS: Pure phosphatidic acids with hydrocarbon chains of 12 and more C-atoms form bilayers in aqueous media.   This is demonstrated for aqueous dispersions of the sodium salt of DLPA. As shown in Figure 1 aqueous dispersions of the phospholipid undergo reproducibly sharp order-disorder transitions on both heating and cooling. The sodium salt of DLPA in the fully hydrated state at pH 7.0 gives a sharp endothermic gel-to-liquid crystal transition at T_m = 32.1 °C with an enthalpy ΔH = 27±1 J/g (= 3.7 kcal/mol). Upon cooling an exothermic transition is observed reproducibly at 28.5°C with an enthalpy ΔH = 28.8 J/g (= 4.0 kcal/mol). The effect of increasing hydration on the thermotropic behavior of sodium DLPA at neutral pH is shown in Figure 2. In the anhydrous state Na^+-DLPA undergoes a crystal-to-liquid crystal transition at 82 °C, with increasing hydration the reversible gel-to-liquid crystal transition temperature decreases continuously reaching a limiting value of ~ 32 °C at water contents greater than 50% (Figure 2). The thermal behavior of aqueous dispersions of synthetic phosphatidic acids with both saturated and unsaturated hydrocarbon chains is summarized in Table I. For comparison the thermal parameters of naturally occuring egg phosphatidic acid are included in the table.

Table I: Transition Temperature (T_m) and Enthalpies (ΔH) of Fully Hydrated Diacylphosphatidates

Phosphatidate	T_m (°C)	ΔH, J/g (kcal/mol)
DLPA	32.1±0.4	27±1 (3.7±0.1)
DMPA	51.9±0.2	39±1 (5.9±0.2)
DPPA	65.3±0.1	53±1 (8.8±0.2)
DSPA	75.2±0.5	59±1 (10.6±0.2)
POPA	28.1±0.1	33±1 (5.7±0.2)
DOPA	-1.9±0.1	32±2 (5.8±0.3)
EPA	21.9±0.8	29±2 (5.0±0.3)

DLPA, 1,2-dilauroyl-sn-glycero-3-phosphoric acid;   DMPA, 1,2-dimyristoyl-sn-glycero-3-phosphoric acid;   DPPA, 1,2-dipalmitoyl-sn-glycero-3-phosphoric acid; DSPA, 1,2-distearoyl-sn-glycero-3-phosphoric acid; POPA, 1-palmitoyl-2-oleoyl-sn-glycero-3-phosphoric acid; DOPA, 1,2-dioleoyl-sn-glycero-3-phosphoric acid; EPA, egg phosphatidic acid. The sodium salt of phosphatidic acids (40-80 mg/ml) were dispersed in 5 mM potassium phosphate buffer pH 7.0 containing 5 mM EDTA. T_m values were derived from heating runs.

On the basis of the thermotropic behavior of Na+-DLPA, X-ray diffraction studies of this phospholipid were carried out as a function of water content at temperatures both below and above the gel-to-liquid crystal transition temperature T_m. Both below and above T_m the low-angle diffraction pattern consisted of lines in the ratio 1: (1/2) : (1/3) : (1/4)....characteristic of a lamellar bilayer phase. Above the transition temperature at 40 °C a single diffuse reflection at 1/4.6 Å^{-1} was observed indicative of a liquid crystalline phase with melted hydrocarbon chains. Below the transition temperature at 20 °C a main sharp reflection is observed at 1/4.2 Å^{-1} indicating the presence of a gel phase with the hydrocarbon chains being packed in a hexagonal lattice. Often a second sharp reflection at 1/4.45 Å^{-1} is observed.

Swelling curves of Na+-DLPA below and above the gel-to-liquid crystal transition temperature are shown in Figure 3A. Below and above the transition temperature the

Table II: Relative fluorescence intensity (%) * of carboxyfluorescein entrapped at self-quenching concentrations (50 mM) in small unilamellar egg phosphatidic acid vesicles.

	Relative fluorescence intensity (%)					
Time(days)	pH-gradient (pH units) **					
	1.0	1.5	2.0	3.2	4.3	5.0
0	20	21	21	21	21	20
2	21	--	21	--	24	24
3	--	30	--	24	--	--
4	23	--	21	--	26	21
5	--	28	--	22	--	--
6	20	--	18	--	20	20
11	--	40	--	26	--	--
12	26	--	20	--	24	74
28	--	56	--	26	--	--
29	54	--	18	--	23	88

* The fluorescence intensity measured when all carboxyfluorescein was released into the bulk phase was taken as 100%.
** A proton gradient was imposed by adjusting the pH of the external medium to alkaline pH. The pH of the internal cavity was kept between pH 6 and 7.

lamellar bilayer repeat distance, d, increases continuously with increasing water content. Such behavior is characteristic of negatively charged phospholipid bilayers. At water contents in excess of 60%, multilayer stacking disorders result in a broadening of the low-angle reflections as indicated by the large error bars (Figure 3A). As shown in Figure 3B, the plot of d vs. (1-c)/c, where c is the concentration of lipid expressed as weight fraction, is linear for both temperatures, indicating that the lipid bilayer structure does not undergo significant structural changes with hydration. Extrapolation of the straight line to (1-c)/c = 0 yields the thickness of the anhydrous Na^+-DLPA bilayer. The values thus obtained are 41 Å and 40 Å at 20 °C and 40 °C, respectively. These values for the bilayer thickness are close to the long or c axis of the unit cell of crystalline monosodium 1,2-dilauroyl-sn-glycerophosphate (Harlos et al.,1984). This phospholipid crystallizes in the monoclinic space group $P2_1/a$ with unit cell axes a = 5.436 Å, b = 7.962 Å, c = 39.04 Å, β = 113.98°.

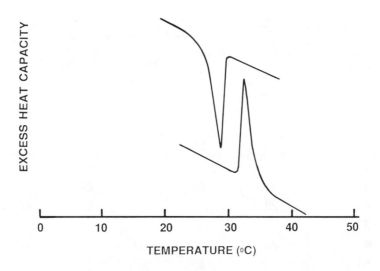

Figure 1: Heating and cooling curves of DLPA (46 mg/ml) dispersed in 5 mM potassium phosphate buffer pH 7.0 containing 5 mM EDTA. Both heating and cooling runs were recorded at 5 °C/min.

If egg phosphatidic acid or DLPA dispersions in water or buffer of neutral pH are subjected to ultrasonication small unilamellar vesicles are produced. Vesicles of similar size and size distribution are obtained by raising the pH of the smectic(lamellar) phosphatidic acid dispersion transiently to values between 10 and 12. This is achieved by either adjusting the pH of the dispersion or alternatively by adding an alkaline solution pH 10 to 12 to the dry film of phosphatidic acid deposited on the glass wall of a round-bottom flask. The alkaline dispersion is then neutralized as quickly as possible to prevent lipid degradation. By this method referred to in the literature as pH-jump or pH-adjustment method (Hauser et al., 1990) and originally described by Hauser & Gains (1982) small unilamellar phosphatidic acid vesicles are formed of an average diameter of 22 nm and a size distribution ranging between 20 and 60 nm. Ion and low-molecular-weight-compounds such as sugars or chromophores, present in the aqueous dispersion medium before the pH is raised, are encapsulated in phosphatidic acid vesicles formed by pH-jump (Gains & Hauser, 1983).

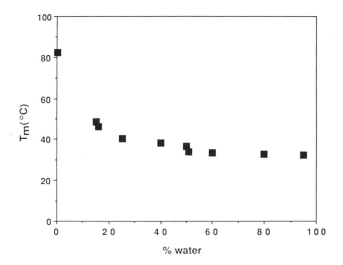

Figure 2: The gel-to-liquid crystal transition temperature T_m of DLPA (46 mg/ml) dispersed in 5 mM potassium phosphate buffer pH 7.0 containing 5 mM EDTA. Transition temperatures T_m were derived from heating curves.

Small unilamellar vesicles of egg phosphatidic acid containing 50 mM carboxyfluorescein in the internal cavity were prepared either by sonication or by pH-jump as described in Materials and Methods. The thermodynamic stability of the resulting vesicles was studied by monitoring the release of carboxyfluorescein as a function of time and pH gradient. The internal pH of the vesicles was taken as the pH

36

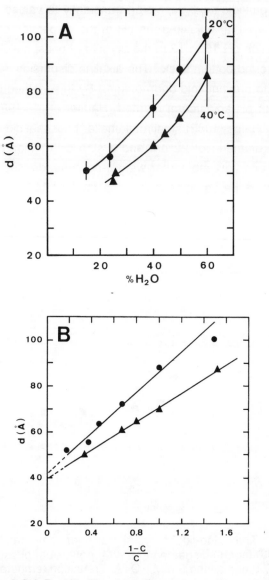

Figure 3 A & B: (A) The lamellar repeat distance, d (Å), of DLPA as a function of the water content (%) below the gel-to-liquid crystal transition temperature T_m (at 20°C, ●) and above T_m (at 40 °C, ▲). The samples were dispersed in H_2O the pH of which was adjusted to neutrality. (B) The lamellar repeat distance, d (Å), of hydrated DLPA was plotted as a function of (1-c)/c where c is the concentration of lipid as weight fraction. The solid lines represent least-squares fits to the experimental data points.

of the original dispersion medium, the external pH was adjusted by gel filtration (see Materials and Methods) such that a pH gradient resulted: the pH of the external medium (outside) was more alkaline than the pH of the vesicle cavity. It should be clear that in these experiments the pH gradient has the same direction as that used to induce vesiculation by pH-jump. From the results summarized in Table II it is concluded that small unilamellar phosphatidic acid vesicles are unstable in the absence of a pH-gradient. In the presence of a pH-gradient of 2-4 pH units the vesicles appear to be stable up to one month. With pH gradients of 5 pH units the external pH is about 12. At alkaline pH phosphatidic acid is known to undergo chemical degradation. As evident from Table II this is accompanied by marked leakage of carboxyfluorescein. The stability of small unilamellar egg phosphatidic acid vesicles produced by pH-jump is demonstrated in Figure 4. After applying a pH

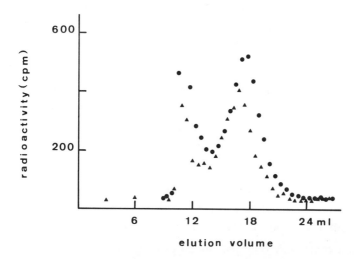

Figure 4: Gel filtration on Sepharose CL-4B (50 x 0.9 cm) of small unilamellar egg phosphatidic acid produced by the pH-jump method (Hauser & Gains, 1982). Egg phosphatidic acid (10 mg/ml) was dispersed in water pH ~3 and the pH was raised to 12 by adding NaOH and returned to pH 9.1 by addition of the appropriate amount of HCl. 0.5 to 1 ml of the dispersion were applied to the column which was equilibrated and run with 0.01 M Tris pH 7.2 containing 0.1 M NaCl, 0.02% NaN3. Elution patterns of egg phosphatidic acid dispersions are shown immediately after preparation (●) and after storage for 30 days at room temperature (▲).

gradient of about 4 units (external medium alkaline) the pH of the external medium was adjusted to pH 9. Using ^{31}P NMR it was shown that the pH of the vesicle cavity was about neutral so that the vesicle stability was measured in the presence of a pH gradient of ~2 pH units at room temperature. Consistent with the data of Table II the vesicle size and size distribution did not change significantly with time. Vesicle size and size distribution were monitored as a function of time by gel filtration on Sepharose CL-4B and freeze-fracture electron microscopy. Elution profiles of the vesicles were recorded as a function of time and the two profiles presented in Figure 4 were recorded after preparation and after storage of the vesicles at room temperature for 1 month.

DISCUSSION: Phosphatidic acids have a tendency to form smectic lamellar phases both in the anhydrous as well as in the fully hydrated state under physiological conditions. This is documented in Figures 1-3 and Table I. Under physiological conditions phosphatidic acids bear one negative charge and the swelling behavior of phosphatidate bilayers in water is characteristic of that observed with negatively charged phospholipids. As a matter of fact the swelling behavior and also the phase behavior of Na$^+$-DLPA is quite similar to that of NH$_4$$^+$-dimyristoylphosphatidylserine (Hauser et al., 1982). The results summarized in Table I indicate that the gel-to-liquid crystal transition temperature of phosphatidates increases with hydrocarbon chain length, and at a given chain length the transition temperature decreases markedly with the number of double bonds. As with other phospholipids the gel-to-liquid crystal transition temperature decreases as the lipid polar group becomes hydrated (Figure 2). A striking feature of the thermal behavior of phosphatidic acids is the high transition temperature T_m (cf. Table I) compared to other negatively charged phospholipids. This becomes obvious if , for instance, the transition temperature T_m of dipalmitoylphosphatidate is compared to that of other phospholipids having the same hydrocarbon chains. T_m of dipalmitoylphosphatidate is 10-15 °C higher than T_m of other negatively charged phospholipids, e.g., it is 10 °C higher than the transition temperature of dipalmitoylphosphatidylserine and about 14 °C higher than T_m of dipalmitoyl-phosphatidylglycerol. T_m of dipalmitoylphosphatidate is, however, close to the transition temperature of isoelectric dipalmitoylphosphatidylethanolamine. Additional intermolecular interaction energies have to be invoked in order to rationalize this finding. The additional interaction energy is very likely to result from hydrogen bonding between the polar groups of phosphatidate molecules (Boggs, 1987).

If smectic (lamellar) phosphatidic acid dispersions are energized, be it in the form of ultrasonic energy or a pH-gradient as with the pH-jump method, small unilamellar vesicles form. Unambiguous evidence has been provided by ^{31}P NMR and infrared spectroscopy to show that the driving force of the pH-jump method is a proton gradient across the phosphatidic acid bilayer (Hauser et al.,1990). Since the formation of small unilamellar phosphatidic acid vesicles of diameter smaller than ~50 nm requires the input of energy, the resulting vesicles are expected to be thermodynamically unstable. This is borne out by experiment. Table II shows that small unilamellar egg phosphatidic acid vesicles are unstable in the absence of a pH-gradient, i.e., if internal and external medium are of similar pH. They are, however, stabilized by a pH-gradient the direction of which is the same as that used to produce vesicles by the pH-jump method. Apparently, as long as a sizeable proton gradient (external pH > internal pH) is present, the highly curved bilayers of small unilamellar phosphatidic acid vesicles are stable and collision-induced fusion as well as leakage of vesicle contents are prevented. It is clear from the data of Table II that as the proton gradient decreases to less than two pH units, vesicles become unstable. However, on the time scale of days even these vesicles appear to be metastable.

REFERENCES

Boggs, J. M. (1987) Biochim. Biophys. Acta 906, 353-404.
Casal, H. L., Mantsch, H. H., Demel, R. A., Paltauf, F., Lipka, G., and Hauser, H. (1990) J. Am. Chem. Soc. 112, 3887 - 3895.
Gains, N., and Hauser, H. (1983) Biochim. Biophys. Acta 731, 31 - 39.
Harlos, K., Eibl, H., Pascher, I., and Sundell, S. (1984) Chem. Phys. Lipids 34, 115 - 126.
Hauser, H., and Gains, N. (1982) Proc. Natl. Acad. Sci. U.S.A. 79, 1683 -1687.
Hauser, H., Paltauf, F., and Shipley, G. G. (1982) Biochemistry 21, 1061 - 1067.
Hauser, H., Mantsch, H. H., and Casal, H. (1990) Biochemistry 29, 2321 - 2329.
Hjelmstad, R. H., and Bell, R. M. (1991) Biochemistry 30, 1731 - 1739.
Schurtenberger, P., and Hauser, H. (1984) Biochim. Biophys. Acta 778, 470 - 480.

Progress in Membrane Biotechnology
Gomez-Fernandez/Chapman/Packer (eds.)
© 1991 Birkhäuser Verlag Basel/Switzerland

TOOLS FOR MOLECULAR GRAPHICS DEPICTIONS OF LIPID STRUCTURES

Bruce Paul Gaber, David C. Turner, Krishnan Namboodiri, William R.Light,II, and Albert Hybl*

Center for Bio/Molecular Science and Engineering, Code 6090, Naval Research Laboratory, Washington, DC 20375, and *Dept.of Biophysics, Univ.of Maryland School of Med., Baltimore, MD 21201

SUMMARY: We have developed several tools for the molecular graphics depiction of lipids and their microstructures. These include: 1) **NanoVision**, a molecular visualization program for Macintosh personal computers; 2) **NRLipid**, a Hypercard database of lipid structural information; 3) **BILAYER BUILDER**, a program for constructing models of lipid bilayers from crystallographic data; and 4) **Ribbon Representation**, an adaptation for the depiction of subtle acyl chain structure of a display tool commonly used for proteins.

NANOVISION

Our initial work on the molecular graphics of lipid structures was conducted on dedicated graphics workstations. These machines, and the software packages which run on them, are unquestionably extremely powerful. However, they have their drawbacks. Computers and software are expensive; out of reach, or at least of limited accessibility for many small laboratories. In part because of their power, dedicated graphics systems are not particularly "user-friendly", often not supporting the range of desktop publishing applications which are becoming increasingly important in scientific communication. With NanoVision we have written a low cost, easy to learn, desktop molecular graphics package for the individual scientist. The program has much of the power of a dedicated system, as well as features not available on dedicated machines. While

NanoVision is a general-purpose molecular graphics program, it was designed with the requirements of lipid depiction very much in mind.

NanoVision is written specifically for the Apple Macintosh line of computers. The Macintosh was chosen because of its powerful graphics-oriented environment, large amount of addressable memory, and ease of use. The program is written in Pascal and meticulously follows established Macintosh programming rules, thus assuring compatibility within the complete line of Macintosh hardware and with all other properly-written applications. As a consequence NanoVision runs on the Mac Plus, Mac SE, and Mac Classic in black and white and in color on the Mac II-series machines. Applications compatibility assures that NanoVision images can be easily transferred via cut and paste to word processors and presentation graphics packages. NanoVision follows the intuitive Macintosh interface guidlines so that it is a program with which a novice or casual user can quickly do molecular graphics. Consequently the program maintains an uncluttered screen and makes extensive use of pull-down menus and mouse control of images (figure 1). Typical display options available with NanoVision include various molecular representations such as skeletal, ball-and-stick, and van der Waals depictions. Multiple windows may be used to display several different structural aspects simultaneously. Images may be rotated and translated with the mouse or by keyboard command. Shading with 256 screen colors are available from a palette of 4 million colors using an exclusive pallet manager.

42

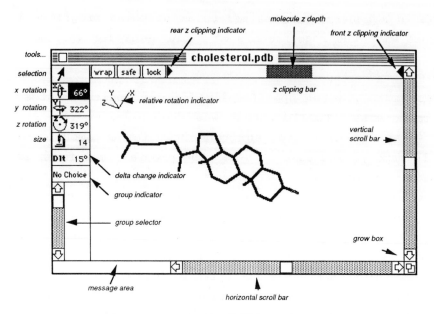

Figure 1: The NanoVision Window

NRLipid

Despite their immense biological and commercial importance, there has been no serious attempt to assemble available structural information about lipids into an easily accessible computer database or knowledge base. NRLipid is an attempt in this direction. NRLipid is an original HyperCard lipid database written in HyperTalk language and implemented on an Apple Macintosh computer. The program permits the crystallographic data related to lipid molecules to be accessible in a computerized filing format. Through mini programs called Scripts the NRLipid database can be used to link other application programs such as NanoVision, WordPerfect and MacWrite to form a central information resource. Using hypercard utilities such as Background, Field, Button, Card and Stack; NRLipid stores, sorts and displays lipid crystallographic data.

HyperCard can create a Background or a template that can be cloned over and over. Information is arranged in Cards which resemble computerized index cards. The user can flip through the Cards to find information. A group of Cards is called a Stack. Thus, NRLipid is a Stack of Cards containing structural information about lipid molecules. Buttons permit linking different Cards or direct the computer to choose an application such as a graphics program. NRLipid stores crystallographic data, links related data, and directs the data to programs for visualization and editing. The prototype NRLipid database currently consists of structural information for a variety of phospholipids and their analogues.

When loaded, the program shows a cover screen that displays the title and authors, the NRLipid button brings the user to the first data card (figure 2). From this point the user can click on either direction arrow to flip forward or backward in the stack. The data in each card can be printed or tabulated into various tables. From this SuperStack card, clicking on the Continued button will call up a dialogue box, asking for a choice between Moreinfo, Picture or None. The selection of Picture will render the crystallographic structure of that particular lipid molecule (figure 2). Similarly, the Locator button will give the choice of Nanovision, Wordperfect or MacWrite. A "Find" command saves blind searching. When used on a Macintosh II in conjunction with NanoVision, lipid structures can be rendered in color.

44

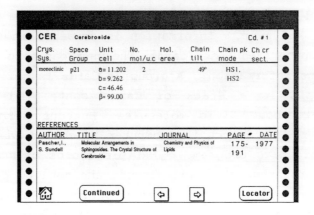

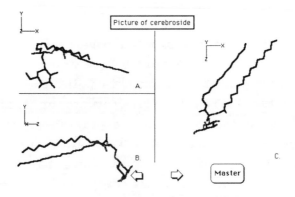

Figure 2: Two Cards from the NRLipid HyperCard stack. The first contains crystallographic data and literature reference for dimyristoylphosphatidylcholine. The second contains four views of the molecule.

BILAYER BUILDER

It is often desirable for non-crystallographers to generate graphical models of three dimensional crystal structures based on published coordinates of the atoms that make up the crystallographic unit cells. This type of visualization is particularly important for studying lipids, where we may be interested in investigating interactions among assembled molecules in addition to the conformations of individual molecules. BILAYER BUILDER is a program which generates a portion of the entire crystal structure from the coordinates of the molecules in a single unit cell. It was written specifically to give users of small desktop computers such as the Apple Macintosh the capability to generate and examine model crystal structures with a molecular graphics display program. In addition, BILAYER BUILDER stores the crystal coordinates in a Protein Data Bank (PDB) file format for possible use in a variety of applications on many different computers.

Lipid crystal structures are reported in the crystallographic literature as the fractional coordinates of a single lipid molecule, or occasionally several molecules, which represent the minimum dataset necessary to reconstruct the crystal. While the coordinates for a single molecule tell us about the configuration of the submolecular groups within the lipid, they are often insufficient to reveal many of the more complex intermolecular interactions among the lipid molecules within the entire lipid assembly. These interactions can be observed by reconstructing the crystal structure from the given atomic coordinates using the space group symmetries and unit cell translational values.

Once constructed, the crystal dataset can be used for many purposes, for example: (1) as input to a graphics display program for visualizing the intermolecular interactions among the lipid molecules, such as hydrocarbon chain packing and interdigitation; or (2) as a starting point for a molecular mechanics or dynamics simulation of a lipid bilayer. The combination of BILAYER BUILDER and NanoVision allows any Macintosh user to investigate the molecular packing in the crystal from many different vantage points. If one is fortunate enough to have access to a more powerful graphics workstation, then BILAYER BUILDER can be combined with a more versatile molecular modelling program such as Quanta (Polygen Corp.) for both visualization and energy calculations.

BILAYER BUILDER is written in ANSI compatible C so that it can be ported to any platform whose compiler supports the ANSI standard. Currently, BILAYER BUILDER runs on any Macintosh with one megabyte RAM and a hard disk drive, and on Silicon Graphics IRIS and Sun SPARC workstations running UNIX. As input, the program requires the magnitudes and angles of the crystallographic unit cell basis vectors and gives the user a choice of 11 space group symmetries covering most of the published space groups for lipid crystals, including sterols and fatty acids. We are continuing to update the program with additional space groups.

Coordinates are read from a standard ASCII file which is compatible with the Brookhaven Protein Data Bank (PDB) format (Bernstein, Koetzle et al. 1977). Since most crystal structures are given in fractional coordinates, the program can perform the required orthogonalization to Cartesian coordinates. The user provides the unit cell dimensions and the space group. There are two steps in hte generation of a structure: (1) completion of the unit cell by operating on the given atomic coordinates with the symmetry operations of the crystal space group, and (2) translation duplication of the completed unit cell. The typical space group symmetry operations include: inversion around a point, mirror reflections, rotations and screw rotations around an axis and glide plane reflections (Mahn 1983).

These operations are used to transform the minimum dataset into a completed unit cell which can then be replicated by translations along the basis vectors (see Figure 3) to generate the model crystal. Where appropriate, the user is allowed to choose among the different possible rotation or screw axes or inversion centers within the crystal. While each of these axes or centers are equivalent in principle, only one choice will generate the most compact configuration of the molecules. For example, figure 3 shows the **a-c** plane of a unit cell with $P2_1$ symmetry in relation to the published coordinates of the A and B molecules of the dimyristoyl phosphatidylcholine (DMPC) crystal (Pearson and Pascher 1979).

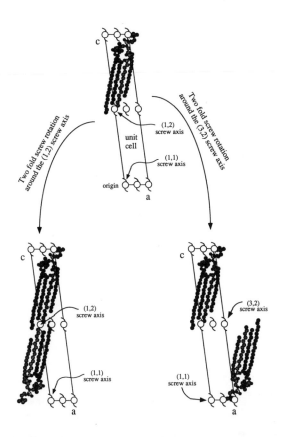

Figure 3: The **a-c** plane of the $P2_1$ monoclinic unit cell of DMPC. The molecules of DMPC-A and B are shown in gray scale. The locations of the two fold screw symmetry axis' are noted, with each labelled such that the axis closest to the origin is the (1,1) axis. The figure shows the screw axis process applied to the two DMPC molecules to make the four molecules that can be translated by lattice vectors to fill the true crystallographic unit cell. The two fold screw rotation around the (1,2) and (3,2) screw axis' is shown for comparison

A portion of four stacked bilayers of Octadecyl-2-methyl-glycerophosphocholine monohydrate (OMPC) (Pascher, Sundell et al. 1986),constructed with BILAYER BUILDER is shown in figure 4. This lipid crystallizes in an orthorhombic unit cell with space group $P2_12_12_1$. Note the total interdigitation of the octadecyl chains and the headgroup-headgroup interaction between successive bilayers.

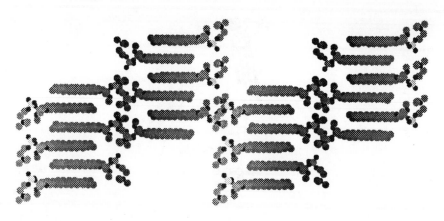

Figure 4 Section along the B-C plane of OMPC constructed using BILAYER BUILDER.

BILAYER BUILDER solves a longstanding problem for membrane biophysicist, that is, how can a lipid crystal structure, or portion of a crystal structure such as a bilayer, be constructed so that it can observed graphically or used as a starting point for a molecular mechanics or dynamics calculation. The ability to propose a trial structure and then easily replicate it opens possibilities for simple starting structures for molecular simulations. The docking of small proteins and peptides into a lipid bilayer is an ideal use of this capability. For example one can make a small patch of bilayer to dock with the peptide and then replicate the proposed structure to an appropriate size for the simulation.

RIBBON REPRESENTATION

The lateral packing of long hydrocarbon chains can be characterized by a pseudo-periodic sub-cell (Abrahamsson, Dahlen et al. 1978). The sub-cell may be triclinic, monoclinic or orthorhombic and the chain-to-chain orientations may be described as parallel or perpendicular. Although this sub-cell nomenclature is useful, it seductively suggests a uniform planar all-trans progression of methylene units along a hydrocarbon chain. Ball-and-stick drawings of the X-ray crystal structure of racemic glycerol 1,2-(di-11-bromoundecanoate) -3- (-p-toluenesulfonate) (DBUTOS) (Watts, Pangborn et al. 1972) suggest a twist in one fatty acid chains and a slight bowing in the other. These are subtleties of lipid structure which are not well depicted with the conventional visual vocabulary of molecular graphics.

Priestle described a suite of FORTRAN programs which use helical ribbons, twisted arrows and smooth ropes (Priestle 1988)for representing the secondary structure of proteins and their topological interrelationships. The ribbon representation captures essential features of protein structure such as a binding cleft or of a protein-substrate interaction. We have adapted this ribbon motif to the depiction of the acyl chains of lipids. In our adaptation, we essentially represent the carbons of fatty acyl chains as if they were peptide C-alpha carbons.

50

Our result for DBUTOS is shown in the stereo illustration in figure 5 he 30 degree twist in chain 1 is clearly illustrated in the ribbon drawing, as is bowing and a slight twist near the terminal end of chain 2. The intercalated parallel packing of the chain 1 terminal ends across the bilayer midline is in sharp contrast to the perpendicular packing of the chain 1 and chain 2 stacks on each side of the bilayer.

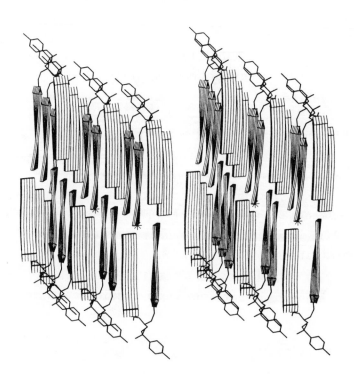

Figure 5:   Stereo pair of the DBUTOS bilayer rendered using a ribbon representation to emphasize the twisting and bowing of the acyl chains.

Acknowledgements: Our work on the development of tools for molecular graphics depictions of lipid structures is supported by the Molecular Biology Program of the Office of Naval Research. We thank Dr. Michael Marron of ONR for his consistent encouragement and support of this project.

REFERENCES

Abrahamsson, S., B. Dahlen, H. Lofgren and I. Pascher. (1978). Prog. Chem. Fats Other Lipids. **16:** 125.

Bernstein, F. C., T. F. Koetzle, G. J. B. Williams, J. Meyer E. F., M. D. Brice, J. R. Rodegers, O. Kennard, T. Shimanouchi and M. Tasumi. (1977). J. Mol. Biol. **112:** 535-542.

Mahn, t. (1983). International Tables for Crystallography. **IV:**

Pascher, I., S. Sundell, H. Eibl and K. Harlos. (1986). Chem. Phys. Lipids. **39:** 53-64.

Pearson, R. M. and I. Pascher. (1979). Nature. **281:** 499-501.

Priestle, J. P. (1988). J. Appl. Cryst. **21:** 572-576.

Watts, J., P. H., W. A. Pangborn and A. Hybl. (1972). Science. **175:** 60-61.

Progress in Membrane Biotechnology
Gomez-Fernandez/Chapman/Packer (eds.)
© 1991 Birkhäuser Verlag Basel/Switzerland

FLUORESCENT ANALOGUES OF PHOSPHOINOSITIDES IN STUDIES ON LIPID-PROTEIN INTERACTIONS AND MEMBRANE DYNAMICS

K.W.A. Wirtz[1], T.W.J. Gadella Jr.[1], J. Verbist[2], P.J. Somerharju[3] and A.J.W.G. Visser[4]

[1]Centre for Biomembranes and Lipid Enzymology, State University of Utrecht, Padualaan 8, NL-3584 CH Utrecht, The Netherlands; [2]Physiological Laboratory, Catholic University of Leuven, Campus Gasthuisberg, Herestraat 49, B-3000 Leuven, Belgium; [3]Department of Medical Chemistry, University of Helsinki, Siltavuorenpenger 10, 00170 Helsinki, Finland; [4]Department of Biochemistry, Agricultural University, Dreijenlaan 3, 6703 HA Wageningen, The Netherlands

INTRODUCTION

Phosphatidylinositol (PI), PI-4-phosphate (PIP) and PI-4,5-bisphosphate (PIP_2) have drawn a great deal of attention mainly because of their key role in agonist-induced transmembrane signal transduction (Berridge and Irvine, 1989; Williamson and Hansen, 1987). Activation of receptors may lead to phospholipase C-dependent degradation of PIP_2 and the formation of inositol 1,4,5-trisphosphate and diglyceride as second messengers. In order to maintain the levels of hormone-sensitive PIP_2 in the plasma membrane, phosphorylation reactions occur that convert PI into PIP_2. PI- and PIP kinases involved in these reactions, are most likely active at the level of the plasma membrane. As a result, the plasma membrane PI-pool will be depleted. Since de novo synthesis of PI does not occur in the plasma membrane (Lundberg and Jergil, 1988 Morris et al., 1990), this pool needs to be replenished by PI from intracellular stores. It has been proposed that the PI-transfer protein may be involved in the intracellular transport of PI to the plasma membrane (Van Paridon et al., 1987a).

Despite their important role in the cell, relatively little is known about the organization and behaviour of PI, PIP and PIP_2 in membranes, or, more specifically, about their tendency to interact with proteins. In order to be able to investigate certain aspects related to these questions, methods have been developed to synthesize fluorescent analogues of PI carrying a parinaroyl-

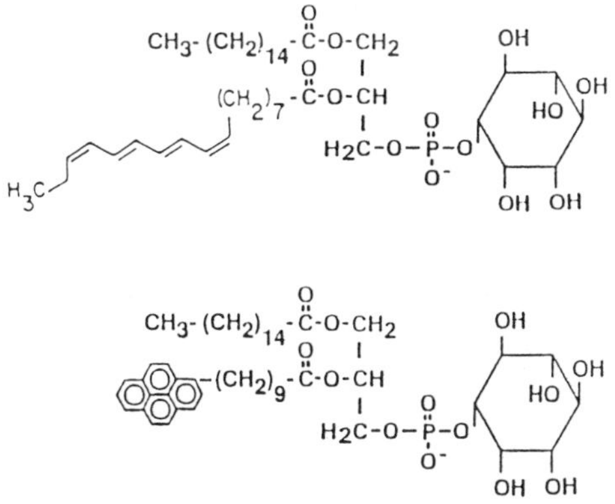

Fig. 1. Structural formulas of 2-parinaroyl-PI and 2-pyrenyl-decanoyl-PI.

or a pyrene-labelled fatty acyl chain in the sn-2-position (Fig. 1) (Somerharju and Wirtz, 1982; Somerharju et al., 1985). More recently, partially purified PI and PIP kinase preparations from bovine brain were found to be very powerful tools in converting pyrenylacyl-labelled PI into the corresponding PIP and PIP_2 analogues (Gadella et al., 1990). Here we will present some studies in which these analogues were used to obtain information on the lipid binding site of the PI-transfer protein, on the dynamic behaviour of phosphoinositides in membranes and on their affinity for the intrinsic membrane protein Ca^{2+}-transporting ATPase.

LIPID BINDING SITE OF THE PI-TRANSFER PROTEIN

PI-transfer protein (PI-TP) belongs to a class of intracellular proteins that is able to bind and transfer phospholipids between membranes (for a recent review, see Wirtz, 1991). The first homogeneous preparation of PI-TP was purified from bovine brain (Helmkamp et al., 1974). Other sources of PI-TP have been bovine heart, rat brain and liver, human platelets and yeast (Helmkamp, 1986). The purified PI-TP's behaved on sodium dodecyl sulphate - polyacrylamide gelelectrophoresis as a protein with a molecular weight in the range of 33,000-35,000. Isoelectric focusing indicated that bovine brain PI-TP consisted of two species (I and II) with isoelectric points of 5.5 and 5.7, respectively (Van Paridon et al., 1987b). This charge difference was due to the fact that PI-TP I contained one molecule of non-covalently bound PI as compared to PI-TP II containing phosphatidylcholine (PC). In agreement with these binding data, it has been well established that PI-TP has a dual specificity with a distinct preference for PI over PC (Demel et al., 1977; DiCorleto et al., 1979; Somerharju et al., 1983).

Application of fluorescent PI and PC analogues which contain a parinaroyl or pyrenylacyl-labelled chain, has enabled us to explore the properties of the lipid binding site of PI-TP (Van Paridon et al., 1987b, 1988a). Time-resolved fluorescence measurements on PI-TP carrying either 2-parinaroyl PI or -PC were carried out to determine the rotational mobility of the 2-parinaroyl chain in the lipid binding site. Analysis of the fluorescence anisotropy decay curves revealed that the dynamic behaviour of the parinaroyl chain was virtually identical for either phospholipid (Fig. 2).

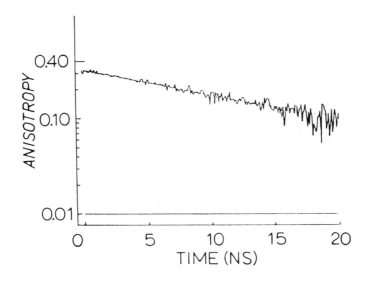

Fig. 2. Fluorescence anisotropy decay of 2-parinaroyl-PI bound to the PI-transfer protein. The experimental anisotropy is represented by the noisy curve, while the smooth line shows the fitted decay function which is a single exponential function. For further details see Van Paridon et al. (1987b).

This decay curve could be represented by a single-exponential function yielding for parinaroyl-PI and parinaroyl-PC rotational correlation times of 16.3 and 17.4 ns, respectively, at 20°C. For comparison, these PI and PC analogues incorporated into PC/PA vesicles (95 : 5, mole%), have correlation times of 1.8 and 2.4 ns, respectively (Van Paridon et al., 1988b). Both the high initial anisotropy (0.32 and 0.31 for PI and PC, respectively) and the relatively long correlation times indicate that the chromophore of the 2-parinaroyl chain of PI and PC are tightly bound to PI-TP. Since these long correlation times are characteristic for the rotation of spherical proteins with a molecular weight of 35,000, we may conclude that the parinaroyl chromophore is completely immobilized in the lipid binding site following the rotational movement of PI-TP.

From the similarity in rotational correlation times, it was argued that the 2-acyl chain of PI and PC may be accommodated in the same hydrophobic pocket on PI-TP (Van Paridon et al., 1987b).

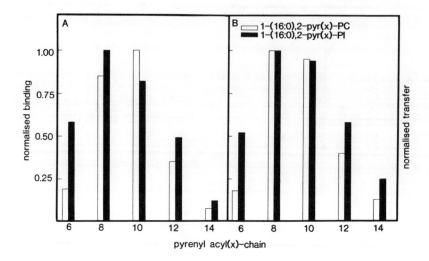

Fig. 3. Comparison between molecular species of PI and PC in terms of binding and transfer by PI-TP. The species carried a palmitoyl (16:0) chain at the sn-1-position and a pyrenylacyl chain of different length at the sn-2-position. The pyrenylacyl chain contained a pyrene moiety at the methyl-terminal end of acyl chains consisting of 6 to 14 C-atoms. For further details, see Van Paridon et al. (1988a).

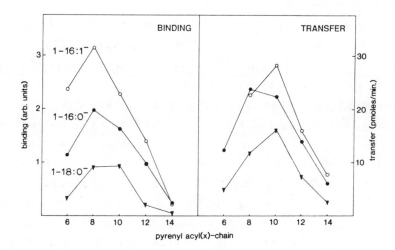

Fig. 4. Effect of acyl chain composition on binding and transfer of PyrPI species by PI-TP. The species carried a palmitoyl (16:0), palmiteoleoyl (16:1) or a stearoyl (18:0) chain in the sn-1 position and a pyrenylacyl chain of different length at the sn-2 position. For further details see Van Paridon et al. (1988a).

This was confirmed by measuring the binding and transfer of PI and PC analogues carrying pyrenylacyl chains of different length (i.e. 6-14 C-atoms) at the 2-position. As shown in Fig. 3, PI-TP has a preference for molecular species which contain a Pyr-C:8 and Pyr-C:10 chain. The similarity in the binding and transfer <u>versus</u> chain length profiles for both sets of species was taken to indicate that the 2-acyl chain of PI and PC share a common binding site (1988a). In addition, by measuring binding and transfer of PI carrying a pyrenylacyl chain at the 2-position and a palmitoyl-(C16:0), palmiteoleoyl- (C16:1) or stearoyl (C18:0) chain at the 1-position, it was shown that PI-TP discriminates between species with the lowest affinity for the C18:0-species (Fig. 4). In analogy what has been found for the PC-transfer protein (Wirtz, 1991) these results strongly suggest that PI-TP has separate binding sites for the 1-and 2-acyl chain.

<u>DYNAMIC BEHAVIOUR IN PC BILAYERS</u>

One of the most striking features of PIP and PIP_2 is their high negative charge content making these phospholipids very acidic at physiological pH. By use of ^{31}P-NMR spectroscopy it could be estimated that PIP is fully ionized (three negative charges) above pH 6.5, and that PIP_2 undergoes a charge shift from -3 to -5 in the pH-range between 6 and 8 (Van Paridon et al., 1986). By way of steady-state fluorescence spectroscopy we have investigated the effect of this high negative charge on the dynamic behaviour of 2-pyrenylacyl-labelled PI (PyrPI), -PIP (PyrPIP) and $-PIP_2$ (PyrPIP$_2$) in organic solvents and in PC vesicle membranes (Gadella et al., 1990). The pyrenyl-chromophore is very well suited for this kind of measurements because upon excitation (for example, at 345 nm) pyrene may give rise to both a monomer and a excimer emission spectrum (Fig. 5). The monomer spectrum is typical for the situation where pyrene in the excited state loses its energy by direct fluorescence emission before having collided with a ground-state

58

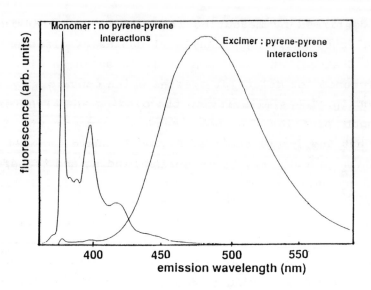

Fig. 5. Monomer and excimer spectrum of PyrPI in PC bilayer.
Wavelength of excitation is 345 nm.

pyrene molecule to form the excited-state dimer or excimer complex. When the excimer complex is formed within the fluorescence
lifetime of the pyrene excited state, it may loose its energy by
fluorescence emission at a longer wavelength resulting in the
characteristic excimer spectrum. The ratio of excimer to monomer
fluorescence (E/M) is proportional to the collision frequency of
the pyrene moieties which is dependent on concentration and
diffusion rate of the probe molecules (Förster, 1969). The
emission spectra of PyrPI, PyrPIP and PyrPIP$_2$, dissolved in
chloroform/methanol are shown in Fig. 6. In contrast to PyrPI that
only displayed a monomer spectrum, both PyrPIP and PyrPIP$_2$
displayed a distinct excimer spectrum indicating that due to their
highly charged state, these phospholipids are poorly soluble and
form small clusters (micelles) in this solvent. When these lipids
were incorporated in a PC vesicle bilayer (present in an amount of
9 mol%) a completely different behaviour was observed. As shown in
Fig. 7, under these conditions PyrPI displayed the most distinct
excimer spectrum. By measuring the fluorescence intensity at 475
nm and 377 nm, one finds that the E/M-value equals 0.69 for Pyr-

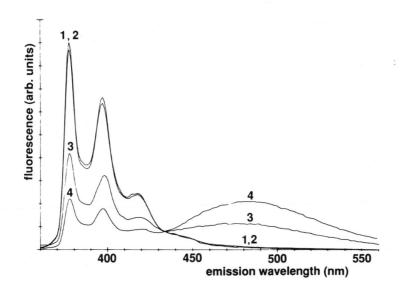

Fig. 6. Fluorescence emission spectra of PyrPC (1), PyrPI (2), PyrPIP (3) and PyrPIP$_2$ (4) in chloroform/methanol (1:1 v/v). For details see Gadella et al. (1990).

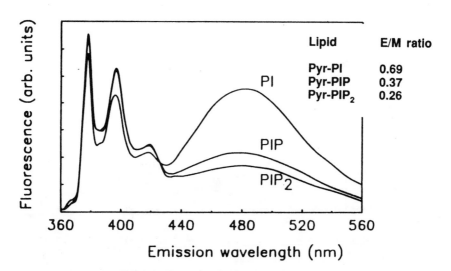

Fig. 7. Fluorescence emission spectra of PyrPI, PyrPIP and PyrPIP$_2$ present in egg PC vesicles. The vesicles contained 9 mole% pyrene phospholipid. For details, see Gadella et al. (1990).

PI, 0.37 for PyrPIP and 0.26 for PyrPIP$_2$. This shows that in PC bilayers the collision frequency decreased in the order PyrPI >

PyrPIP > PyrPIP$_2$ as a result of increased charge repulsion between molecules. Ca^{2+}-levels had to be increased to above 0.1 mM to observe an increase in E/M for PyrPIP$_2$ (9 mole%) in PC vesicles. In agreement with the observations of Toner et al. (1988) relatively high concentrations of calcium are required to have an effective shielding of the negative charge on PIP$_2$. Calcium barely has an effect on the collision frequency of PyrPIP and no effect on the diffusion behaviour of PyrPI.

AFFINITY FOR (Ca^{2+}+Mg^{2+})-ATPase

In a previous study (Missiaen et al., 1989) it was shown that the activity of the plasma-membrane Ca^{2+}-transporting ATPase purified from smooth muscle and from erythrocytes, was stimulated by phosphoinositides in the order PIP$_2$ > PIP > PI. In order to determine the interaction of these phosphoinositides with the ATPase, the enzyme was reconstituted in a mixture of PC and the

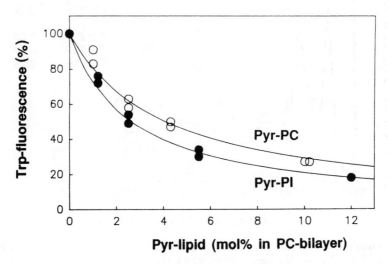

Fig. 8. Quenching of tryptophan-fluorescence of reconstituted (Ca^{2+}+Mg^{2+})-ATPase by PyrPC (o) and PyrPI (o) as a function of the pyrene-lipid concentration in egg-PC bilayers. For details see Verbist et al. (1991).

pyrene-labelled analogues, followed by measuring fluorescence
energy transfer between the tryptophanyl residues and the pyrene
moieties present (Verbist et al., 1991). These measurements are
possible as the tryptophanyl emission spectrum of $(Ca^{2+}+Mg^{2+})$-
ATPase (excitation at 290 nm) overlaps with the pyrene-absorption
spectrum.

The efficiency of fluorescence energy transfer was determined
by measuring the quenching of the tryptophan fluorescence of the
$(Ca^{2+}+Mg^{2+})$-ATPase   as   a   function   of   the   concentration   of
zwitterionic PyrPC and anionic PyrPI in the reconstituted system
(Fig. 8). At concentrations above 10 mol% acceptor lipid,
tryptophan fluorescence was reduced by about 80%. Furthermore it
was evident that PyrPI was more efficient in quenching the
tryptophan fluorescence than PyrPC possibly reflecting a higher
affinity of the $(Ca^{2+}+Mg^{2+})$-ATPase for the negatively charged
phospholipid. The effect of 1 mole% PyrPI, PyrPIP and $PyrPIP_2$ on
the tryptophanyl fluorescence emission of the reconstituted ATPase
is shown in Fig. 9. Fluorescence quenching due to energy transfer

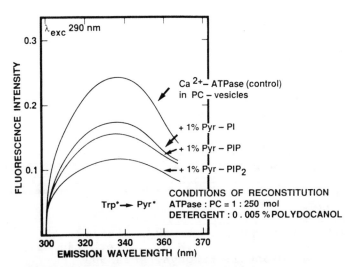

Fig. 9. Fluorescence emission spectra of purified $(Ca^{2+}+Mg^{2+})$-
ATPase reconstituted in egg PC (control) and in egg PC that
contains 1 mole% PyrPI, PyrPIP and $PyrPIP_2$. For details, see
Verbist et al. (1991).

62

increased in the order PyrPIP$_2$ > PyrPIP > PyrPI. This strongly suggests that (Ca^{2+}+Mg^{2+})-ATPase attracts PIP$_2$ more efficiently than PIP or PI which may explain the effective stimulation of ATPase activity by PIP$_2$ in the reconstituted system (Missiaen et al., 1989). This is the first direct evidence in support of the model that in the plasma membrane PIP$_2$ may not be homogeneously distributed in the inner aspect of the bilayer but rather may be concentrated around transmembrane proteins including ion pumps and receptors.

REFERENCES

Berridge, M.J. and Irvine, R.F. (1989) Nature 341, 197-205.
Demel, R.A., Kalsbeek, R., Wirtz, K.W.A. and Van Deenen, L.L.M. (1977) Biochim. Biophys. Acta 466, 10-22.
DiCorleto, P.E., Warach, J.B. and Zilversmit, D.B. (1979) J. Biol. Chem. 254, 7795-7802.
Förster, T. (1969) Angew. Chem. 81, 364-374.
Gadella, T.W.J., Moritz, A., Westerman, J. and Wirtz, K.W.A. (1990) Biochemistry 29, 3389-3395.
Helmkamp, G.M., Harvey, M.S., Wirtz, K.W.A. and Van Deenen, L.L.M. (1974) J. Biol. Chem. 249, 6382-6389.
Helmkamp, G.M. (1986) J. Bioenerg. Biomembr. 18, 71-91.
Lundberg, G.A. and Jergil, B. (1988) FEBS Lett. 240, 171-176.
Missiaen, L,, Raeymaekers, L., Wuytack, F., Vrolix, M., De Smedt, H. and Casteels, R. (1989) Biochem. J. 263, 687-694.
Morris, S.J., Cook, H.W., Byers, D.M., Spence, M.W. and Palmer, F.B.St.C. (1990) Biochim. Biophys. Acta 1022, 339-347.
Somerharju, P.J. and Wirtz, K.W.A. (1982) Chem. Phys. Lipids 30, 81-91.
Somerharju, P., Van Paridon, P. and Wirtz, K.W.A. (1983) Biochim. Biophys. Acta 731, 186-195.
Somerharju, P.J., Virtanen, J.A,, Eklund, K.K., Vaino, P. and Kinnunen. P.K.J. (1985) Biochemistry 24, 2773-2781.
Toner M, Vaio G, McLaughlin A, McLaughlin S (1988) Biochemistry 27:7435-7443.
Van Paridon, P.A., De Kruijff, B., Ouwerkerk, R. and Wirtz, K.W.A. (1986) Biochim. Biophys. Acta 877, 216-219.
Van Paridon, P.A., Gadella, T.W.J., Somerharju, P.J. and Wirtz, K.W.A. (1987a) Biochim. Biophys. Acta 903, 68-77.
Van Paridon, P.A., Visser, A.J.W.G. and Wirtz, K.W.A. (1987b) Biochim. Biophys. Acta 898, 172-180.
Van Paridon, P.A., Gadella, T.W.J., Somerharju, P.J. and Wirtz, K.W.A. (1988a) Biochemistry 27, 6208-6214.
Van Paridon, P.A., Shute, J.K., Wirtz, K.W.A. and Visser, A.J.W.G. (1988b) Eur. Biophys. J. 16, 53-63.

Verbist, J., Gadella, T.W.J., Raeymaekers, L., Wuytack, F., Wirtz, K.W.A. and Casteels, R. (1991) Biochim Biophys Acta <u>1063</u>, 1-6.
Williamson, J.R., Hansen, C.A. (1987) In: Litwack G (ed) Biochemical Action of Hormones, vol. 14. Academic Press, New York, pp. 29-80
Wirtz, K.W.A. (1991) Ann. Rev. Biochem. <u>60</u>, 73-99.

64

Progress in Membrane Biotechnology
Gomez-Fernandez/Chapman/Packer (eds.)
© 1991 Birkhäuser Verlag Basel/Switzerland

BINDING, INTERACTION, AND ORGANIZATION OF PROTEINS WITH LIPID MODEL MEMBRANES

D. W. Grainger, K. M. Maloney, X. Huang
Biomolecular Materials Research Center, Oregon Graduate
Institute of Science and Technology, Beaverton, OR 97006-1999
USA

M. Ahlers, A. Reichert, H. Ringsdorf
Institut für Organische Chemie, Universität Mainz, D6500 Mainz,
Germany

C. Salesse
Centre de Recherche en Photobiophysique, Universite a Trois-
Rivieres, Trois-Rivieres, Quebec G9A 5H7 Canada

J. N. Herron, V. Hlady, K. Lim
Center for Biopolymers at Interfaces, University of Utah, Salt
Lake City, UT 84112 USA

SUMMARY: Model membrane systems are used to investigate
protein recognition and binding at interfaces. Fluorescence
microscopy results are presented for interactions of the proteins,
phospholipase A_2 and antifluorescyl IgG, at lipid monolayer
interfaces. Total internal reflection fluorescence measurements
are used to quantify albumin and IgG adsorption to supported
lipid monolayers.

INTRODUCTION

In the assembly and function of the cell membrane, proteins interact with the lipid bilayer by binding or adsorption to its surface, insertion into the hydrocarbon region of the membrane interior, or penetration through the membrane, forming a lipid-protein alloy. Lipid chemistry and resulting microstructures in the membrane that influence protein interaction and the recognition of membrane features are aspects of biomembrane architecture that are important to many new technologies. Studies of the natural membrane paradigm may well prove advantageous both in fundamental biophysical terms to elucidate how lipid-membrane-protein association results in diversity of function, but also to understand how natural membrane surfaces are tailored to promote specific protein interactions and eliminate nonspecific events.

This communication reports a number of simplified model biomembrane systems that address aspects of protein recognition and binding to model membrane systems. The model membranes simplify an otherwise difficult investigative scenario using complex plasma membranes, plus have encouraged the notion that through such systems, one can manipulate interfacial properties to induce desired responses in artificial ways that may prove useful to technology (Ahlers et al., 1990d).

MATERIAL AND METHODS

Video-Enhanced Epifluorescence Microscopy of Lipid Monolayers and Lipid-Protein Interactions at the Air-Water Interface:
Recently fluorescence microscopes have been configured to allow direct visualization of monolayer lipid films at the air-water interface; domains of organized amphiphilic molecules formed by a phase transition within the monolayer film from liquid-expanded to solid-condensed physical states are normally observed (Meller, 1988; Lösche & Möhwald, 1984; Weis et al., 1984; Weis & McConnell, 1984). Analogously, this technique has also proven valuable to study interactions of labeled proteins with monolayers (Blankenburg et al., 1989; Ahlers et al., 1990a, 1990b, 1990d; Grainger et al., 1989; 1990a). Protein labeled with a fluorescent marker is introduced into the subphase under a lipid monolayer and its interaction with the layer monitored both visually over time and as a function of lipid physical state (varying surface pressure). A specially designed miniaturized, thermostatted Langmuir film balance on the stage of an epifluorescence microscope limits the required quantities of protein to microgram

scales (Meller, 1989).

Phospholipase A$_2$-Phospholipid Monolayer Studies: Various phospholipids were spread as pure monolayers and compressed into their phase transition regions, providing a physically heterogeneous monolayer surface comprising fluid lipid coexisting with domains of solid phase organized lipid. Fluorescein-labeled phospholipase A$_2$ (PLA$_2$, Naja naja) was introduced under these monolayers and the hydrolytic reaction of the enzyme against the monolayer followed under the microscope.

Cationic Dye Binding Studies: Cationic water-soluble fluorescent dye, 1,1',3,3,3',3'-hexamethylindocarbocyanine iodide (Molecular Probes, Eugene, OR) was dissolved in buffer (0.4 μM concentration). Ternary mixed monolayer systems of DPPC, palmitic acid, and lysopalmitoylphosphatidylcholine (Avanti Polar Lipids, Birmingham, AL) containing 1 mol% lipid dye were spread from chloroform solutions on buffer subphases at 30°C. Various ratios of DPPC to equimolar concentrations of palmitic acid and lysoPC (e.g., 1:3:3 DPPC:palmitic acid:lysoPC, 1:5:5, 1:1:1, etc.) were examined. These monolayers were compressed under the fluorescent microscope until substantial monolayer phase separation was observed (typically 25-30 mN/m surface pressure). Cationic dye solution was then carefully injected into the subphase underneath the phase-separated monolayer and observed through the rhodamine filter.

Binding and Quenching of Fluorescent Lipid Haptens by a Monoclonal Antifluorescyl Antibody: Various fluorescent lipids containing fluorescein as the headgroup (shown in Fig. 1) were used as lipid-bound haptens to investigate the interfacial binding requirements of antibodies on biomembrane surfaces (Ahlers et al., 1990b, 1990c). A murine monoclonal antibody produced against fluorescein has been well characterized in terms of its affinity for fluorescein (Kranz & Voss, 1981; Kranz et al., 1982; Herron, 1984; Herron et al., 1986; Gibson et al., 1988) and, most recently, its high resolution crystal structure has been elucidated (Herron et al., 1989). This (the 4-4-20 clone), IgG$_{2a}$(k) antibody binds fluorescein with a binding constant of 3.4 x 10^{10} M^{-1} in

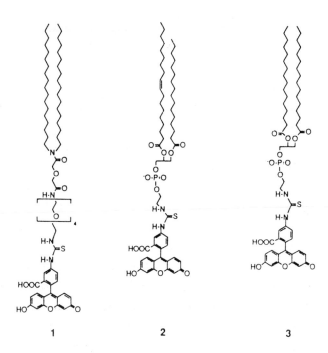

Fig. 1: Structure of fluoresceinated lipids. (1) DODA-(EO)$_4$-FITC; (2) egg PE-FITC; (3) DPPE-FITC.

aqueous solution (Bates et al., 1985), quenches fluorescence upon binding up to 96% (Kranz & Voss, 1981), and shows no cross reactivity with rhodamine (Kranz & Voss, 1981; Voss et al., 1976), making the fluorescein lipid-antibody binding pair a unique and highly suited model for these studies. A portion of this monoclonal antibody was labeled with rhodamine X isothiocyanate (XRITC, Molecular Probes). Quenching reactions as well as interfacial protein domain formation could be observed by fluorescent microscopy of these lipid monolayers.

Investigations on the binding of antifluorescyl antibody to fluorescent lipids also relied on data from fluorescent mixed micelles and vesicles (Ahlers et al., 1990b). Micelle fluorescence in buffer was titrated with antibody and kinetics of fluorescent quenching were monitored over time. Giant lipid vesicles containing phospholipids mixed with fluorescent lipids were swollen from lipid films in a thermostatted flow chamber placed under an inverted fluorescence microscope (Decher et al., 1990). Antibody was introduced to quench lipid fluorescence signal in the outer vesicle membrane leaflets for fluorescent lipids of varying chemical architectures.

Experimental data were verified using computer-driven molecular dynamics simulation models (Ahlers et al., 1990b). The recent crystallographic structure of the monoclonal antibody binding domain (Fv fragment) (Herron et al., 1989) was included with a matrix of lipid headgroups in a monolayer. One fluorescent lipid was included in the center of the lipid matrix and the antibody binding domain was "bound" to this single lipid by computer manipulation. Subsequently, the entire model was subject to energy minimization iteratively until reasonable energy minima were achieved. Molecular dynamics were run under various conditions simulating the aqueous environment of the experiments (48 picoseconds duration) during which the interaction energies and structural changes in the model were monitored.

Total Internal Reflection Fluorescence (TIRF) Studies: Purified human immunoglobulin (IgG) and human albumin (HSA) labeled with fluorescein isothiocyanate (FITC) were used in these studies. Labeling efficiencies were approximately 2 for IgG and 0.7 for HSA. Proteins were dissolved in phosphate buffer (PBS, pH 7.3, I = 0.19). FITC-IgG concentration was 0.050 mg/ml; FITC-HSA concentration was 0.041 mg/ml.

TIRF Instrumentation: A system consisting of a TIRF flow cell, an argon ion laser, fluorescence collection optics, a monochromator, and a charge-coupled device (CCD) camera described previously (Hlady, 1991) was used to image and quantitate adsorption of proteins to supported lipid monolayers. The CCD is capable of recording changes in fluorescence intensity simultaneously along one linear dimension of a sample surface over time. Excitation light is totally internally reflected through the quartz prism optically coupled to the silica plates supporting the lipid membranes in the flow cell. Emission signal from labeled protein within the evanescent field across 2 cm of the silica plate surface i detected by the CCD camera. For each experiment, protein was adsorbed for 11 min from a flowing solution (flow rate 0.5 ml/min) and subsequently desorbed under buffer flow at the same flow rate.

Supported Lipid Monolayers: Organized lipid films were transferred from a commercial film balance (KSV, Helsinki, Finland) to the surfaces of freshly silanized amorphous silica slides (ESCO, New Jersey) hydrophobized with octadecyltrichloro-silane (Thompson et al., 1984). Supported monolayers of dihexadecylphosphate (DHP), dimyristoylphosphatidylcholine (DMPC), and dioctadecyldimethylammonium bromide (DODAB) were transferred to supports from pure water (>18 Mohm resistivity) at 20°C at surface pressures of 35-40 mN/m (45-55 Å^2/molecule). Films were transferred on the downstroke, yielding a supported layer with lipid headgroups exposed to the aqueous phase. Supports containing films were captured in clean glass vials under the subphase, covered, and transported to the TIRF instrument under water. Additionally, some lipid layers incorporated 1 mol% of fluorescein-labeled egg lecithin (Avanti Lipids, Alabama) for tests of layer stability under flow cell shear.

RESULTS AND DISCUSSION

Phospholipase A$_2$ Hydrolysis of Phospholipid Monolayers and Consequent Enzyme Domain Formation: PLA$_2$ is responsible for catalyzing hydrolysis of the 2-acyl ester bond of 3-sn-glycerophospholipids to yield the acyl fatty acid and corresponding lysolipid. Observed activation of this enzyme in response to organized interfacial phospholipid substrates as opposed to dispersed lipid monomers has made it an ideal investigative tool for probing membrane structure as well as model enzymatic behavior.

We have documented the progress typical of enzyme recognition, binding, hydrolysis, and ultimately domain formation in several phospholipid monolayers (Grainger et al., 1989, 1990a, 1990b). Large enzyme domains (up to 100 μm long) form in the lipid monolayer as lipid is hydrolyzed. The disappearance of solid

phase lipid domains is accompanied by assembly of protein
aggregates that grow in size and number as hydrolysis proceeds.
Enzyme aggregation occurs well before most of the monolayer is
hydrolyzed. Whenever hydrolysis occurs, the same pattern of
hydrolytic behavior--binding, hydrolysis, and PLA$_2$ domain
formation--is observed consistently in phospholipid monolayers.

A mechanism for the formation of organized, two-dimensional
enzyme domains within monolayers of phospholipids is proposed in
Fig. 2. Active enzyme under the layer recognizes its substrate,

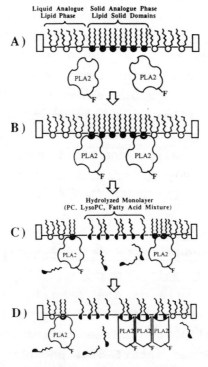

Fig. 2: Proposed mechanism for recognition, binding, hydrolysis
and subsequent PLA$_2$ domain formation prompted by critical
concentrations of hydrolytic endproducts mixed with substrate
lipids in phospholipid monolayers. (A) Injection of PLA$_2$ into
aqueous subphase under lipid monolayer. (B) Recognition and
binding of PLA$_2$ to lipid interface. (C) Hydrolysis of monolayer
by PLA$_2$ with subsequent buildup of hydrolytic products (fatty
acid and lysolipid) within the layer. (D) Organization of bound
PLA$_2$ into protein domains at the lipid-water interface prompted
by critical concentrations and phase separation of hydrolytic
endproducts in the monolayer (Ahlers et al., 1990b).

binds to the monolayer and hydrolyzes in an interfacial region between liquid and solid lipid phases. After a critical extent of hydrolysis, products of hydrolysis--lysolipids and fatty acids-- build up in localized regions of the layer. Phase separation of at least the fatty acid from pure lipid occurs, leading to areas of increased charge density in the case of fatty acids. In vesicle systems, such fatty acid phase separation and charge densities have already been shown (Yu et al., 1989). Enzyme, with its cationic surface charge then binds and builds domains in these areas of localized negative charge, leading to the phenomena of enzyme domain formation witnessed in DPPC, DMPC, and DPPE monolayers (Grainger et al., 1989, 1990a).

Cationic Dye Binding to Phase-Separated Domains in Mixed Monolayers: We have recently attempted to unequivocally demonstrate fatty acid phase separation in monolayers using a cationic dye and fluorescent microscopy. Fig. 3 shows results of

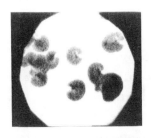
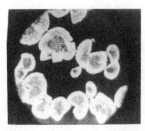

Rhodamine Filter
Dark Domains

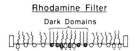

- Mixed monolayers of PC, lysoPC, and fatty acid are fluid
- Phase separation of membrane
- Domains visualized by membrane packing differences

Cationic dye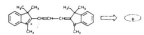

Fluorescein Filter
Dark Domains

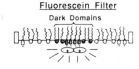

- Cationic dye binds selectively to anionic membrane domains
- Visualize docking of dye to domains

Fig. 3: Electrostatic binding of water-soluble cationic dye, 1,1',3,3,3',3'-hexamethylindocarbocyanine iodide, to anionic phase-separated domains in ternary mixed monolayers containing phospholipid, lysolipid, and fatty acid (1:5:5 mole ratio) at 30°C, surface pressure 27 mN/m, subphase 100 mM NaCl, 10 mM Tris, 5 mM CaCl$_2$, pH 8.0.

a cationic fluorescent dye binding to ternary mixed monolayers of DPPC, palmitic acid, and lysoPC. These monolayers have been compressed to a point where fluorescence microscopy can detect large grey domains. These domains do not exhibit the black opaque character of classic crystalline lipid phases formed during a first-order fluid-solid phase transition, and their grey character has been attributed to phase separation of fatty acid components within the monolayer (Ahlers et al., 1990d). Many different stoichiometric mixtures of DPPC:fatty acid:lysolipid demonstrate these domains and we have produced grey domains with a strikingly similar morphology to the PLA_2 domains seen upon hydrolysis. Data from zwitterionic vesicle systems (Yu et al., 1989) shows strong adsorption of cationic dye to vesicles after buildup of fatty acid during PLA_2 hydrolysis. Our rationale was meant to show a relationship between monolayer fatty acid domains (negative surface charge) analogous to those created by PLA_2 hydrolysis of phospholipid monolayers, and cationic dye (PLA_2 model) by electrostatic binding.

As shown in Fig. 3, grey phase-separated areas seen before cationic dye addition transform to bright fluorescing domains after dye addition, demonstrating a rapid and stable electrostatic adsorption of dye to selected domains at the interface. Dye is not adsorbed if no fatty acid is present in monolayer mixtures. Use of buffer of pH 4 or elimination of calcium in the buffer (using EDTA) both abolish the phase separation and dye binding, indicating the fatty acid content of phase-separated areas. These results present strong evidence in line with our hypothesis (Fig. 2) that phase-separated domains are, in fact, high in fatty acid concentration. Although the pK_a of the fatty acid at such an interface is elevated to near 7.0, a high negative surface charge density above pH 8.0 rapidly binds cationic dye, creating maximum observed fluorescent signal seen on the domains in contrast to the rest of the mixed monolayer. This has much relevance to the mechanism of PLA_2 domain formation described above, namely that PLA_2, with its concentration of cationic (lysine) residues clustered on the enzyme's surface near its binding site (White et al., 1991; Scott et al., 1991), readily adsorbs electrostatically to phase-separated fatty acid domains of negative charge in the monolayer after hydrolysis produces enough fatty acid to induce phase separation. This electrostatic adsorption process is also what has been proposed to activate PLA_2 on phosphatidylcholine vesicles (Yu et al., 1989). This dye study proves that phase-separated areas of fatty acid are highly charged and act to bind dye and/or enzyme to yield large fluorescent domains.

Binding and Quenching of Fluorescent Lipid Haptens at Biomembrane Interfaces by a Monoclonal Antifluorescyl Antibody: Various chemistries for the presentation of the fluorescein hapten are available (Ahlers et al., 1990c; Petrossian et al., 1985; Struck & Pagano, 1980; Knight & Stephens, 1989) and its incorporation into various model biomembrane systems is rather simple (Struck & Pagano, 1980) so that physicochemical factors governing the binding of fluorescein by anti-fluorescein antibodies may be systematically studied.

Regardless of the physical nature of the interface presented to the antibody (micelle vs. vesicle vs. monolayer), antibody recognition of the fluorescent hapten measured by quenching remains dependent only on lipid chemistry. Fig. 4 shows data for the fluorescence quenching behavior of the different fluorescein-lipids by the antibody as observed in monolayers. Quenching varies from $\approx 70\%$ for DODA-(EO)$_4$-FITC, to $\approx 50\%$ for DPPE-FITC, and to $\sim 10\%$ for egg PE-FITC at 30 mN/m. These absolute values are also very similar to those obtained in the fluorescence quenching experiments in octyl glucoside micelles as well as in giant vesicles (Ahlers et al., 1990b). Measurements with the XRITC-labeled antibody show that black domains which form in the monolayer by lipid quenching using the fluorescein filter were fluorescent when excited by the rhodamine filter (Fig. 5). This indicates directly that the dark monolayer aggregates are large clusters of rhodamine-labeled antibodies bound to quenched fluorescein-lipids at the monolayer interface.

74

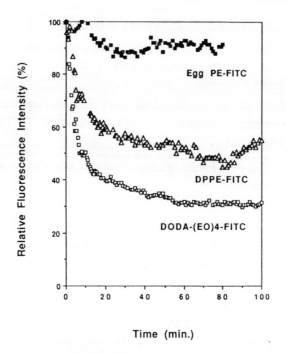

Fig. 4: Extent of fluorescence quenching by 4-4-20 antibody of mixed monolayers containing 1 mol% DODA-(EO)$_4$-FITC, DPPE-FITC, or egg PE-FITC in 99 mol% POPC at 30 mN/m. Spreading solvent: CHCl$_3$. Subphase is phosphate buffer (50 mM, pH 8.0, 20°C).

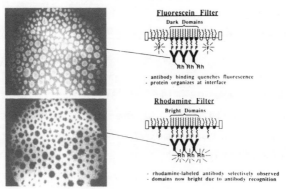

Fig. 5: Two-dimensional protein domains of 4-4-20 anti-fluorescein antibody bound to monolayers of 1 mol% DODA-(EO)$_4$-FITC in 99 mol% POPC at 12 mN/m and 20°C as observed by fluorescence microscopy of the monolayer interface. Scale bar is 50 μm. Upper micrograph: taken using a fluorescein filter (monolayer signal); lower micrograph: taken using a rhodamine filter (labeled antibody signal).

Quantitation of fluorescein hapten recognition and quenching is complicated by bilayer leaflets: only the outer leaflet exposed to extra-vesicular aqueous medium is accessible to the antibody. Quenching of fluorescence occurs, therefore, only on the outer leaflet, and can approach a value of approximately 65% (Petrossian & Owicki, 1984). Because quenching of fluorescein-lipids on the outer membrane leaflet cannot be visually distinguished from fluorescence emission from the unquenched inner leaflet using direct fluorescence observation of giant vesicles, the XRITC-labeled antibody was used for detecting binding. Using a qualitative photometric measurement of antibody rhodamine signal, the amount of antibody found to bind on vesicles of the various lipids was found to correlate directly with results from both micelles and monolayers: magnitude of response was recognition of DODA-$(EO)_4$-FITC > DPPE-FITC >> Egg PE-FITC (Ahlers et al., 1990b).

Complete reviews of the theory and methods of molecular simulations of biological molecules have been published (Brooks et al., 1988; McCammon & Harvey, 1987; Hagler, 1985). Molecular dynamics simulations reported for this system assume that binding has occurred between Fv and fluorescein hapten at the start of the MD run. Based on the initial coordinates of fluorescein binding in the binding site (based on crystallographic data), fluorescein and Fv structures are allowed to relax within the constraints of the force field approximations. Two types of data are drawn from these simulations (Ahlers et al., 1990b): 1) a distance-time dependence between amino acid residues in the binding pocket and the fluorescein hapten in the pocket (interactive changes in fluorescein binding), and 2) an RMS deviation calculation of $C\alpha$ atoms on the Fv backbone changing as a function of MD run time (protein structural assessment).

Crystallography of the Fv-fluorescein complex shows that the binding pocket is long and narrow--a cleft between the Fv light and heavy chains (Herron et al., 1989). In binding fluorescein, a light chain arginine residue deep in the binding pocket in a low dielectric medium (ARG39L) is hypothesized to interact strongly with the xanthoyl ring of fluorescein, forming an ion pair. Thus, the distance-time relationship of the ARG39L residue with a xanthoyl enolate oxygen has been calculated over the duration of the 48 picosecond MD run.

Another result of the MD runs compares the RMS deviations in amplitude of the distance-time molecular trajectories between ARG39L and fluorescein (Ahlers et al., 1990b). Far greater fluctuations in amplitude of motion are present for Fv-fluor and Fv-DODA than for Fv-DPPE. The interaction between ARG39L and fluorescein seems to be hindered or dampened--one clue that the lack of a spacer system immobilizing fluorescein to the lipid hinders its free movement and natural configurational fluctuations. Both free fluorescein and DODA-(EO)$_4$-FITC allow the fluorescein hapten freedom to move within the Fv binding domain.

The covalently conjugated fluorescein hapten must penetrate the binding site sufficiently close to ARG39L to form a salt bridge with an enolic oxygen of fluorescein's xanthene ring. Insufficient penetration would result in only partial complex formation, decreased binding affinity, and partial or no quenching. 15 Å is needed as a spacer length to allow an immobilized, covalently attached fluorescein hapten full ability to enter the antibody active site (Ahlers et al., 1990b). Only DODA-(EO)$_4$-FITC is able to fulfill this requirement experimentally. The long, hydrophilic spacer connecting fluorescein to the lipid has an approximate length of 23.3 Å, based on an energy-minimized CPK model. DPPE-FITC and Egg PE-FITC clearly lack a spacer, as the length from the isothiocyanate of fluorescein to the phosphodiester group amounts to only a fraction of the required length (6.9 Å). The MD evidence clearly shows the hindered approach of the Fv fragment to bind DPPE-FITC at the monolayer interface, as well as the dramatic conformational relaxation required to accommodate such binding. Fig. 6 shows the differences in the approaches by the Fv 4-4-20 fragment to the monolayer models containing either of the fluoresceinated lipids, DODA-(EO)$_4$-FITC and DPPE-FITC. It

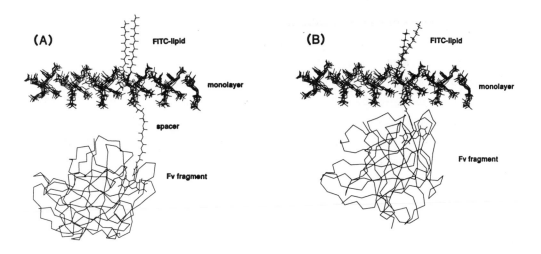

Fig. 6: Comparisons of the Fv fragment crystal structure (Cα-
backbone) of the 4-4-20 antibody fully bound to fluorescein
lipids, and energy-minimized for 48 picoseconds in the molecular
dynamics lipid monolayer model. Perspective is viewing the
monolayer side-on and is identical in both pictures. (A) Fv-
DODA-(EO)$_4$-FITC complex at the model lipid monolayer interface; (B) Fv-DPPE-FITC complex at the model lipid monolayer
interface. The fluorescein xanthoyl ring is discernible within the
Fv binding cleft in both models.

is readily apparent that DODA's long spacer (Fig. 6A) allows an
easy approach of the antibody to bind the fluorescein hapten
without any steric interference from the monolayer interface
(Ahlers et al., 1990b). Alternatively, DPPE-FITC (Fig. 6B)
presents the fluorescein hapten so close to the interface that, in
order to gain full binding for quenching, the Fv fragment must
actually penetrate the monolayer. This steric effect must cause
conformational changes in the protein and disrupt monolayer
packing. Thus, less fluorescein in this system is quenched
experimentally.

TIRF on Supported Monolayers: Interfacial detection of adsorbed
proteins using total internal reflection evanescent wave optics
has been utilized for the past 25 years (for reviews see Axelrod
et al., 1984; Hlady et al., 1985). TIRF techniques can access

several experimental variables: the amount of adsorbed protein and protein adsorption kinetics, the conformational and orientational states of adsorbed protein, and the interaction of adsorbed protein with ligands (enzyme-substrate, antibody-antigen). Recent developments in the TIRF technique (Hlady, 1991) have combined a CCD-based photon detection system to yield spatially resolved, fast protein adsorption and desorption kinetics at interfaces.

We are interested in the technique to monitor plasma protein interactions with lipid membrane systems. Recent observations reported for liposomal membranes (Allen et al., 1991; Chonn et al., 1990) and blood contacting surfaces (Hayward & Chapman, 1984; Hall et al., 1989) demonstrate some interesting interactions between lipid surfaces and blood components. However, quantification of protein interactions with lipid membrane surfaces is not available and the effects of lipid composition and physical states in membranes on protein adsorption and binding are poorly understood. The potential to mimic biomembrane surface structure to yield useful, practical biomaterial surfaces is clearly indicated.

Supported lipid monolayers containing fluorescein lipids were used to test the stability of the lipid films under shear in the flow cell. Under the flow regime used in this study (shear = 100-200 cm^{-1}), fluorescent intensities monitored from the lipid layer were photobleached over time (decreased fluorescence) but flow of buffer did not change the decay kinetics due to bleaching. This was taken as evidence of the stabilty of these lipid membranes during flow studies.

Preliminary results of HSA adsorption to DHP and DMPC supported monolayers are shown in Fig. 7. Data analysis is based on two assumptions: namely, that fluorescence signal detected scales linearly with the amount of adsorbed protein (fluorescence quantum yield does not change with time); and that adsorption is transport-limited in all regions and the steepest portion of each curve equals the transport-limited rate (Hlady, 1991). Adsorption amounts in each case are calculated using the Leveque equation.

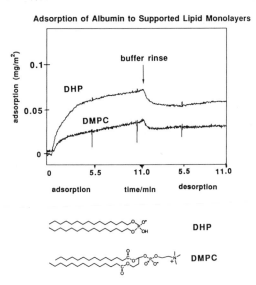

Fig. 7: TIRF results for adsorption of labeled albumin on various supported lipid monolayers.

The curves in Fig. 7 are characterized by an initial adsorption, a plateau region (although in these cases not quite to equilibrium), a desorption profile leading to a second plateau region which yields irreversible adsorbed amounts of HSA on the lipid films. Adsorbed amounts after buffer flush are greater for negatively charged DHP surfaces (0.06 mg/m^2) than for zwitterionic DMPC surfaces (0.04mg/m^2). In neither case is a monolayer of protein adsorbed. In fact, comparisons with other surfaces indicate that relatively little HSA is adsorbed on these lipid surfaces. HSA adsorption onto hydrophilic quartz surfaces has been quantified to be 0.11 mg/m^2. Additionally, HSA adsorption to poly(ethyleneoxide) derivatized surfaces has been recorded at 0.07 mg/m^2--the lowest adsorption levels of any surface tested to date (Hlady, unpublished). One can ascertain, therefore, that lipid surfaces suppress the irreversible adsorption of HSA in the TIRF system.

We also have results for IgG adsorption to DHP, DMPC, and positively charged DODAB surfaces. This case shows that adsorption of IgG to lipid surfaces is higher than that seen for HSA. Monolayers or near-monolayers of IgG are produced on all surfaces. Unlike HSA, DHP adsorbs less IgG than DMPC (0.42 mg/m^2 vs. 0.49 mg/m^2), while DODAB adsorbs the least amount. These results indicate 10 times as many IgG molecules are adsorbed to the surfaces compared to HSA after rinsing.

While these results are a first attempt to quantitate protein adsorption in situ to lipid membranes, many more experiments must be run before a great amount of information on these systems is available. However, the ability to extend these investigations to mixed lipid layers of any composition is enhanced by the use of Langmuir-Blodgett technology. Additionally, the use of polymerizable lipids (gift of Biocompatibles, Ltd., England) to stabilize these supported layers under handling and and varied flow and shear conditions is currently underway.

Acknowledgements: D.W.G. was supported by a Postdoctoral Research Fellowship from the Alexander von Humboldt Foundation, West Germany and acknowledges partial support from the Basic Energy Sciences Program, U.S. Department of Energy. C.S. was supported by a Postdoctoral Research Fellowship from the Natural Sciences and Engineering Council of Canada. Research in Mainz is assisted by a BMFT Grant for Ultrathin Films (03M4008F1). This work was partially supported by Public Health Service Grant #Al 22898 (JNH) and the Center for Biopolymers at Interfaces, University of Utah (J.N.H, V.H.). J. N. Herron and K. Lim thank Biosym Technologies for the generous donation of DISCOVER and INSIGHT molecular graphics programs and Silicon Graphics Inc. for the SGI 4D/70GT workstation.

REFERENCES

Ahlers, M., Blankenburg, R., Grainger, D.W., Meller, P., Ringsdorf, H., and Salesse, C. (1990a) Thin Solid Films 180, 93.
Ahlers, M., Grainger, D.W., Herron, J.N., Lim, K., Ringsdorf, H., and Salesse, C. (1990b) submitted to Biophys. J.
Ahlers, M., Grainger, D.W., Ringsdorf, H., and Salesse, C. (1990c), submitted to Bioconj. Chem.
Ahlers, M., Müller, W., Reichert, A., Ringsdorf, H., and Venzmer, J. (1990d) Angew. Chem. Int. Ed. Engl. 29, 1269.
Allen, T.M., Austin, G.A., Chonn, A., Lin, L., and Lee, K.C. (1991) Biochim. Biophys. Acta 1061, 56.
Axelrod, D., Burghardt, T.P., and Thompson, N.T. (1984) Annu. Rev. Biophys. Bioeng. 13, 247.
Bates, R.M., Ballard, D.W., and Voss, E.W., Jr. (1985) Mol. Immunol. 22, 871.

Blankenburg, R., Meller, P., Ringsdorf, H., and Salesse, C. (1989) Biochemistry 28, 8214.

Brooks, C.L., Karplus, B.M., and Pettit, B.M. (1988) Adv. Chem. Phys. 71, 1.

Chonn, A., Lafleur, M., Kitson, N., Devine, D.V., and Cullis, P.R. (1990) 10th Int. Biophysics Cong., Vancouver, BC, Canada.

Decher, G., Ringsdorf, H., Venzmer, J., Bitter-Suermann, D., and Weisgerber, C. (1990) Biochim. Biophys. Acta 1023, 357.

Gibson, A.L., Herron, J.N., He, X.-M., Patrick, V.A., Mason, M.L., Lin, J.-N., Kranz, D.M., Voss, E.W., Jr., and Edmondson, A.B. (1988) Proteins, 3, 155.

Grainger, D.W., Reichert, A., Ringsdorf, H., and Salesse, C. (1989) FEBS Lett. 252, 73.

Grainger, D.W., Reichert, A., Ringsdorf, H., and Salesse, C. (1990a) Biochim. Biophys. Acta, 1023, 365.

Grainger, D.W., Reichert, A., Ringsdorf, H., Salesse, C., Davies, D., and Lloyd, J.B. (1990b) Biochim. Biophys. Acta 1022, 146.

Hagler, A.T. (1985) The Peptides 7, 213.

Hall, B., Bird, M., Kojima, M., and Chapman, D. (1989) Biomaterials 10, 219-224.

Hayward, J.A., and Chapman, D. (1984) Biomaterials 5, 135-142.

Herron, J.N. (1984) In: Fluorescein Hapten: An Immunological Probe (E.W. Voss, Jr., Ed), CRC Press, Boca Raton, FL, pp. 49-76.

Herron, J.N., He, X.-M., Mason, M.L., Voss, E.W., Jr., and Edmundson, A.B. (1989) Proteins 5, 271.

Herron, J.N., Kranz, D.M., Jameson, D.M., and Voss, E.W., Jr. (1986) Biochemistry 25, 4602.

Hlady, V. (1991) Appl. Spectrosc. 45, 246-252.

Hlady, V., Van Wagenen, R.A., Andrade, J.D. (1985) In: Surface and Interfacial Aspects of Biomedical Polymers, vol. 2 (J.D. Andrade, Ed), Plenum Press, New York, pp. 81-119.

Knight, C.G., and Stephens, T. (1989) Biochem. J. 258, 683.

Kranz, D.M., Herron, J.N., and Voss, E.W., Jr. (1982) J. Biol. Chem. 257, 6987.

Kranz, D.M., and Voss, E.W., Jr. (1981) Mol. Immunol. 18, 889.

Lösche, M., and Möhwald, H. (1984) Rev. Sci. Instrum. 55, 1968.

McCammon, J.A. and Harvey, S.C. (1987) Dynamics of Proteins and Nucleic Acids, Cambridge Univ. Press, Great Britain.

Meller, P. (1988) Rev. Sci. Instrum. 59, 2225.

Meller, P. (1989) J. Microsc. Oxford 156, 241.

Petrossian, A., Kantor, A.B., and Owicki, J.C. (1985) J. Lipid Res. 26, 767.

Petrossian, A., and Owicki, J.C. (1984) Biochim. Biophys. Acta 776, 217.

Scott, D.S., White, S.P., Otwinoski, Z., Yuan, W., Gelb, M.H., and Sigler, P.B. (1991) Science 250, 1541.

Struck, D.K., and Pagano, R.E. (1980) J. Biol. Chem. 255, 5404.

Thompson, N.T., Biran, A., McConnell,H.M. (1984) Biochim. Biophys. Acta 722, 10.

Voss, E.W., Jr., Eschefeld, W., and Root, R.T. (1976) Immunochemistry, 13, 447.

Weis, R.M., and McConnell, H.M. (1984) Nature 310, 47.

82

Weis, R.M., Tamm, L.K., and McConnell, H.M. (1984) Proc. Natl.
 Acad. Sci. U.S.A. <u>81</u>, 3249.
White, S. P., Scott, D. L., Otwinowski, Z., Gelb, M. H., and
 Sigler, P. B. (1991) Science <u>250</u>, 1560.
Yu, B.Z., Kozubek, Z., and Jain, M.K. (1989) Biochim. Biophys.
 Acta <u>980</u>, 23.

Progress in Membrane Biotechnology
Gomez-Fernandez/Chapman/Packer (eds.)
© 1991 Birkhäuser Verlag Basel/Switzerland

TRANSPORT, EQUILIBRIUM, AND KINETICS OF ANTIBODIES ON PLANAR
MODEL MEMBRANES MEASURED BY QUANTITATIVE FLUORESCENCE MICROSCOPY

Nancy L. Thompson, Mary Lee Pisarchick, Claudia L. Poglitsch,
Melanie M. Timbs, Martina T. Sumner and Helen V. Hsieh

Department of Chemistry, University of North Carolina, Chapel
Hill, North Carolina, USA

SUMMARY: Several aspects of the behavior of antibodies at
substrate-supported planar membranes have been characterized with
quantitative fluorescence microscopy: (1) very slow rotational
mobilities of an anti-dinitrophenyl monoclonal antibody associated
with solid-like phospholipid Langmuir-Blodgett monolayers have
been measured with polarized fluorescence photobleaching recovery;
(2) the equilibrium and kinetics of an anti-dinitrophenyl Fab at
phospholipid Langmuir-Blodgett monolayers have been characterized
with total internal reflection fluorescence microscopy; and (3)
the apparent association constants of a panel of anti-
dinitrophenyl mouse IgG1 antibodies with substrate-supported
planar membranes containing purified and reconstituted mouse IgG
Fc receptors have been measured with total internal reflection
fluorescence microscopy.

A key step in antibody-mediated immune responses is the
association of antibodies with cell surfaces. Parameters which
may play a role in antibody-mediated cellular functions and which
are not yet well understood include the thermodynamic and kinetic
properties of antibody-membrane association, the translational and
rotational   mobilities   of   membrane-bound   antibodies,   the
interaction of membrane-bound antibodies with other cell surface
or circulating molecules, and the conformations of membrane-
associated antibodies.   An in-depth understanding of the
association of antibodies with membrane surfaces is also important

in the development and application of biosensors and immunodiagnostic devices. This paper describes the use of planar model membranes deposited on optically transparent surfaces (McConnell et al., 1986) with techniques in dynamic fluorescence microscopy (Thompson et al., 1988) to investigate antibody-membrane association.

ROTATIONAL MOBILITY OF ANTIBODIES ON LANGMUIR-BLODGETT MONOLAYERS

Information about molecular size and shape, molecule-membrane association, and membrane viscosity is reflected by the rate and degree of the rotational mobility of membrane-bound species. Very slow rotational motions of a fluorescent lipid in solid-like phospholipid Langmuir-Blodgett monolayers and of fluorescently labeled anti-dinitrophenyl (DNP) antibodies specifically bound to the monolayers have been measured with polarized fluorescence photobleaching recovery (P-FPR; Smith et al., 1981; Velez and Axelrod, 1988; Timbs and Thompson, 1990).

Monolayers were composed of distearoylphosphatidylcholine (DSPC) (100 mol%) or a mixture of dinitrophenyldioleoylphosphatidyl-ethanolamine (DNP-DOPE) (30 mol%) and DSPC (70 mol%). Using a Joyce-Loebl Langmuir Trough (Model 4) at room temperature, monolayers were spread at 100 $Å^2$/molecule, compressed to 35 dyn/cm, and deposited onto octadecyltrichlorosilane-treated glass microscope slides. Monolayers contained 0.5 mol% 1,1'-dioctadecyl-3,3,3',3'-tetramethylindocarbocyanine (diI) and/or were treated with the mouse IgG1 anti-dinitrophenyl monoclonal antibody ANO2 (Balakrishnan et al., 1982).

ANO2 antibodies were purified from hybridoma supernatants using protein A affinity chromatography. For measurements of antibody mobility, ANO2 antibodies were conjugated with the bifunctional, amine-reactive, tetramethylindocarbocyanine fluorophore CY3.18 (C-). This fluorophore may covalently react at two sites and therefore might experience less independent flexibility than unifunctional probes (Wagner et al., 1990). All antibodies were clarified by 0.22 μm filtration and sedimentation at 100,000 g for 2 hours before use with monolayers.

For microscopy, monolayers were washed with phosphate-buffered saline (PBS; 0.05 M sodium phosphate, 0.15 M sodium chloride, 0.01% sodium azide, pH 7.4), and treated with sheep IgG in PBS (100 μg/ml, 2 min) to block nonspecific binding sites. Monolayers were then treated with ANO2 or C-ANO2 in PBS (concentration and time as specified), and washed with PBS to remove unbound antibodies. The presence and mobility of diI and C-ANO2 were examined using a fluorescence microscope constructed from an inverted optical microscope (Zeiss IM-35) and an argon ion laser (Coherent Innova 90-3).

The specificity of C-ANO2 for DNP-DOPE in the monolayers was demonstrated by the following results (Timbs and Thompson, 1990): C-ANO2 binds to DNP-DOPE/DSPC monolayers in a saturating manner; C-ANO2 does not bind significantly to DSPC monolayers; and C-ANO2 binding to DNP-DOPE/DSPC monolayers is blocked by excess DNP-glycine (DNP-G) or unlabeled ANO2. C-ANO2 remains bound for at least two hours.

DiI and C-ANO2 lateral mobilities were measured with fluorescence pattern photobleaching recovery (Smith and McConnell, 1978; Timbs et al., 1991) using circularly polarized light and a spatial periodicity of 4 μm. These measurements indicated that the lateral mobilities of both diI and C-ANO2 were slow ($\leq$ 35% mobile with a diffusion coefficient $\leq$ 2 x 10^{-10} cm^2/sec). The expected fractional fluorescence recovery arising from lateral motion, for the large illuminated area used with P-FPR (25 μm radius for diI and 38 μm radius for C-ANO2), was less than 0.002 (Axelrod et al., 1976).

Rotational mobility was measured with P-FPR using linearly polarized light (Smith et al., 1981; Velez and Axelrod, 1988). P-FPR anisotropies r(t) are defined as

$$r(t) = \frac{\Delta F_{par}(t) - \Delta F_{per}(t)}{\Delta F_{par}(t) + 2\Delta F_{per}(t)} \tag{1}$$

where

$$\Delta F_{par,per}(t) = F(-) - F_{par,per}(t) \tag{2}$$

$F_{par}(t)$ and $F_{per}(t)$ are the post-bleach fluorescence intensities

86

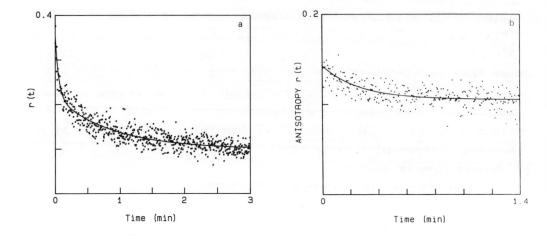

Figure 1. **P-FPR Anisotropies**. Typical measured r(t) for (a) diI in DSPC monolayers and (b) C-ANO2 on DNP-DOPE/DSPC monolayers decrease with time. The lines are the best fits to Eq. 3 with D_1 = 0 and (a) n = 3 or (b) n = 2. Parameters were as follows: laser wavelength, 514.5 nm; observation and bleaching powers, 3-30 μW and $\leq$ 0.4 W; bleaching duration, 50 milliseconds (diI) or 2 seconds (C-ANO2); bleaching depth, 15-35%; temperature, 18 $^\circ$C.

measured with light polarized parallel and perpendicular to the bleach polarization, respectively, and F(-) is the pre-bleach fluorescence. For azimuthally symmetric, two-dimensional samples in which n different fluorophore populations rotationally diffuse with coefficients D_i (Velez and Axelrod, 1988; Timbs and Thompson, 1990),

$$r(t) \quad = \quad \frac{2\,(b/a)\,\sum\limits_{i=1}^{n} f_i\,\exp(-4D_i t)}{3\,-\,(b/a)\,\sum\limits_{i=1}^{n} f_i\,\exp(-4D_i t)} \quad (3)$$

where the f_i are the relative abundances of the different fluorophore types. Constants a and b depend on the bleaching depth and the orientation distribution and flexibility of the absorption and emission dipoles and are assumed to be equal for the different fluorophore populations. These constants are also the average fractional bleach and recovery, i.e.,

TABLE I: PARAMETERS OF DiI AND C-ANO2 ROTATIONAL MOBILITY

Parameter	DiI	C-ANO2	C-ANO2 and Anti-IgG
a	0.25 ± 0.02	0.27 ± 0.03	0.27 ± 0.04
b	0.09 ± 0.02	0.05 ± 0.01	0.04 ± 0.01
$(4D_2)^{-1}$ (sec)	99 ± 11	72 ± 12	84 ± 65
$(4D_3)^{-1}$ (sec)	4.5 ± 0.7	-	-
f_1	0.46 ± 0.12	0.73 ± 0.17	0.94 ± 0.24
f_2	0.24 ± 0.04	0.27 ± 0.04	0.06 ± 0.03
f_3	0.30 ± 0.06	-	-

Values are the means and standard errors in the means obtained from parameters of the best fits of approximately 100 P-FPR recovery and decay curves to Eq. 3 with n = 3 (diI) or n = 2 (C-ANO2) and D_1 = 0.

$$a = [F_{per}(0) + F_{par}(0)]/[2F(-)] \qquad (4)$$
$$b = [F_{per}(0) - F_{par}(0)]/[2F(-)] \qquad (5)$$

Measured P-FPR anisotropies r(t) for diI in DSPC and DNP-DOPE/DSPC monolayers, and for C-ANO2 on DNP-DOPE/DSPC monolayers, decayed with characteristic times of 1-100 seconds and were nonzero at longer times (Figure 1). The anisotropy data were fit to Eq. 3, assuming that the fluorophores were fully mobile ($D_i \neq$ 0 for i > 1) or immobile (D_1 = 0), and with n = 1, 2, 3 or 4. Analysis with an F-statistic (Wright et al., 1988) indicated that the diI and C-ANO2 data were best described by two (n = 3) and one (n = 2) mobile component(s), respectively, and an additional immobile component.

The r(t) for diI were not statistically different for DSPC monolayers, DNP-DOPE/DSPC monolayers, or for DNP-DOPE/DSPC monolayers treated with unlabeled ANO2 (20 μg/ml, 10 min). Also, the r(t) for C-ANO2 on DNP-DOPE/DSPC monolayers did not change for several sets of conditions with which monolayers were treated with C-ANO2 (20 μg/ml, 10 min; 50 μg/ml, 10 min; 20 μg/ml, 20 hr). However, crosslinking monolayer-bound C-ANO2 (20 μg/ml, 10 min) with polyclonal anti-(mouse IgG) antibodies (100 μg/ml, 10 min) did significantly decrease the rotationally mobile fraction. The constants a and b, the average rotational correlation times $(4D_i)^{-1}$, and the fractional abundances f_i for diI and C-ANO2 data are shown in Table I.

The very slow rotational correlation times for diI are consistent with previous measurements of diI rotational mobility in solid-phase multibilayers and liposomes (Smith et al., 1981; Johnson and Garland, 1983). Several phenomena might give rise to the experimentally observed non-monophasic anisotropies for diI and C-ANO2, including restricted rotational motion (Kinosita et al., 1977), different lipid environments (McConnell et al., 1984), and the presence of lipid and/or antibody aggregates (Ethier et al., 1983; Uzgiris, 1983).

EQUILIBRIUM AND KINETICS OF Fabs ON LANGMUIR-BLODGETT MONOLAYERS

To fully understand the mechanisms underlying antibody-mediated processes, it is necessary to characterize the thermodynamic and kinetic aspects of antibody-membrane associations. Therefore, the equilibrium and kinetics of fluorescently labeled ANO2 Fab at phospholipid Langmuir-Blodgett monolayers containing DNP-conjugated phospholipids have been examined with total internal reflection fluorescence microscopy (TIRFM; Axelrod et al., 1984).

Solid-like monolayers composed of dipalmitoylphosphatidylcholine (DPPC) (100 mol%) or a mixture of dinitrophenylaminocaproylphos-phatidylethanolamine (DNP-cap-DPPE) (25 mol%) and DPPC (75 mol%) were deposited on alkylated fused quartz slides as described above. ANO2 antibodies were purified from hybridoma supernatants using affinity chromatography with DNP-conjugated human serum albumin (DNP-HSA) and ANO2 Fabs were produced, purified and labeled with fluorescein isothiocyanate (F-) as described (Pisarchick and Thompson, 1990). All antibodies were clarified by 0.22 μm filtration and sedimentation at 134,000 g and 22 °C for 30 minutes before use with monolayers.

Monolayers were washed with PBS, treated with sheep IgG/PBS (1 mg/ml, 5 min) and then with F-(ANO2 Fab) (concentration as specified, $\geq$ 15 min) in 1 mg/ml sheep IgG/PBS. Under these

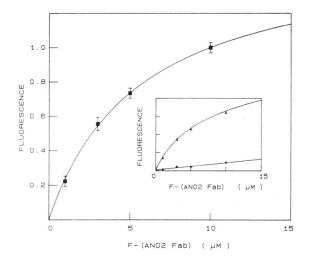

Figure 2. F-(ANO2 Fab) on Phospholipid Monolayers. The difference (■) of the evanescently excited fluorescence of F-(ANO2 Fab) on DNP-cap-DPPE/DPPC (▲) and DPPC (♦) monolayers increases in a saturating manner with the F-(ANO2 Fab) solution concentration. The best fit of the difference data to Eq. 6 (——) gives K = 2 x 10^5 M^{-1}. Analysis with Eq. 7 gives [N] ≈ 6000 molec/μm^2. Uncertainties are standard errors in the means for averages over nine spatially independent measurements on each of three monolayers. Parameters for TIRFM were as follows: illuminated area, approximately 50 μm x 200 μm; laser wavelength, 488.0 nm; laser power, 30 μW; temperature, 25-28 $^{\circ}$C.

conditions, the F-(ANO2 Fab) specifically binds to the DNP-cap-DPPE, the adsorption process reaches a steady-state, and the F-(ANO2 Fab) solution concentrations adjacent to the monolayers are equal to the applied concentrations (Pisarchick and Thompson, 1990).

In TIRFM, the thin evanescent field (depth d ≈ 850 Å) created by a totally internally reflected laser beam selectively excites membrane-bound fluorescent molecules (Poglitsch and Thompson, 1990). Using this technique, the surface-associated fluorescence of F-(ANO2 Fab) on DNP-cap-DPPE/DPPC and DPPC monolayers was measured as a function of the F-(ANO2 Fab) solution concentration. The fluorescence for DNP-cap-DPPE/DPPC monolayers was high and began to saturate, whereas the fluorescence on DPPC monolayers was significantly lower and linear with Fab concentration (Figure 2).

The fluorescence on DPPC monolayers was attributed to F-ANO2 Fab

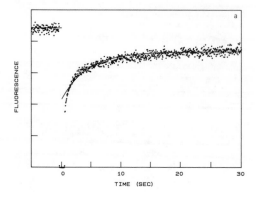 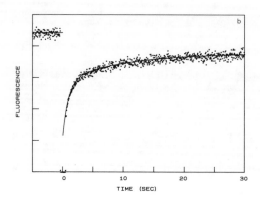

TIME (SEC)   TIME (SEC)

Figure 3. <u>TIR-FPR Recovery Curves</u>. Shown is a typical recovery curve for F-(ANO2 Fab) on DNP-cap-DPPE/DPPC monolayers and the best fits to Eq. 8 with $k_1 = 0$ and (a) n = 2 or (b) n = 3. Parameters were as follows: [F-(ANO2 Fab)], 3 μM; laser wavelength, 488.0 nm; observation and bleaching powers, 30 μW and $\leq$ 0.5 W; bleaching duration, 50-200 ms; bleaching depth, 30-80%; temperature, 25-28 °C.

within the evanescent field but not bound to the surface, and was subtracted from the fluorescence on DNP-cap-DPPE/DPPC monolayers. The difference data were fit to the shape for a bimolecular reversible reaction between monovalent Fab and surface sites, or

$$F([A]) = \frac{F(\infty) \; K \; [A]}{1 + K \; [A]} \qquad (6)$$

where [A] is the solution concentration of fluorescent ligand, K is the apparent membrane association constant, and $F(\infty)$ is the fluorescence at saturation. The ratios R of the fluorescence intensities on DNP-cap-DPPE/DPPC monolayers to those on DPPC monolayers gave estimates of the surface density of bound fluorescent molecules at saturation, [N], according to the expression (Poglitsch et al., 1991)

$$[N] = (R - 1)d[A]/\beta \qquad (7)$$

TABLE II: TIR-FPR RATES AND FRACTIONAL RECOVERIES

[F-(ANO2 Fab] (μM)	1	10	10
Area (μm x μm)	50 x 200	50 x 200	200 x 550
k_2 (sec^{-1})	0.11 $\pm$ 0.01	0.11 $\pm$ 0.01	0.10 $\pm$ 0.01
k_3 (sec^{-1})	1.22 $\pm$ 0.16	1.23 $\pm$ 0.04	1.21 $\pm$ 0.05
f_1	0.17 $\pm$ 0.04	0.18 $\pm$ 0.01	0.17 $\pm$ 0.01
f_2	0.35 $\pm$ 0.03	0.30 $\pm$ 0.01	0.29 $\pm$ 0.01
f_3	0.48 $\pm$ 0.02	0.52 $\pm$ 0.01	0.54 $\pm$ 0.01

Values are the means and standard errors in the means obtained from best fits to Eq. 8 (with n = 3 and k_1 = 0) of 27 to 45 TIR-FPR recovery curves from three to five independently prepared monolayers.

where d is the evanescent depth and β is the fractional surface site saturation for a given solution concentration of fluorescent ligand.

Total internal reflection illumination was combined with fluorescence photobleaching recovery (TIR-FPR) to obtain information about surface binding kinetic rates and surface diffusion coefficients (Thompson et al., 1981). TIR-FPR recovery curves for ANO2 Fab on DNP-cap-DPPE/DPPC monolayers, obtained as described (Burghardt and Axelrod, 1981), had characteristic times in the range of seconds but were not monoexponential and did not completely recover (Figure 3). The data were therefore analyzed with the functional form

$$F(t)/F(-) = 1 - \sum_{i=1}^{n} f_i \exp(-k_i t) \qquad (8)$$

with k_1 = 0. Analysis with an F-statistic (Wright et al., 1988) indicated that the fits were significantly better for n = 3 (as compared to n = 2) but not for n = 4 (as compared to n = 3).

The rates and fractional recoveries for the best fits to Eq. 8 with n = 3 are shown in Table II. These parameters did not change as a function of the Fab solution concentration, indicating that the recovery curves were determined primarily by the intrinsic kinetics of the surface reaction rather than the rate of solution transport. In addition, the parameters did not change when the illuminated and bleached area was increased approximately 10-fold in size, indicating that surface diffusion did not significantly contribute to the rate and shape of the fluorescence recovery. TIR-FPR curves were also independent of the bleaching depth.

The non-monoexponential nature of the TIR-FPR data indicates that the association of F-(ANO2 Fab) with DNP-cap-DPPE/DPPC monolayers is more complex than a simple bimolecular surface reaction. Possible explanations include heterogeneous surface binding sites (McConnell et al., 1984), heterogeneous antibody solution conformations (Theriault et al., 1991), and multistep Fab-DNP reactions (Herron et al., 1986).

ANTIBODIES ON PLANAR MEMBRANES CONTAINING moFcγRII

The combination of TIRFM and functional reconstitution of membrane receptors in supported planar membranes provides a means of characterizing the thermodynamics and kinetics of weak ligand-receptor interactions. In this work, the association of three monoclonal mouse IgG1 anti-DNP antibodies and polyclonal mouse IgG with the mouse IgG Fc receptor moFcγRII reconstituted into substrate-supported planar membranes has been characterized with TIRFM. The association was examined in the absence and presence of saturating amounts of DNP-G to investigate the potential effect of bound monovalent hapten on IgG-moFcγRII interactions.

MoFcγRII is a glycosylated, single transmembrane polypeptide belonging to the immunoglobulin supergene family. This receptor binds monomeric IgG1 and IgG2b with relatively low affinity (Burton, 1985) and is found on a number of immunological cell types, including macrophages and lymphocytes (Unkeless et al., 1988; Mellman et al., 1988). Ligation of the macrophage moFcγRII with multivalent IgG can induce several cellular immune responses including receptor-mediated endocytosis of immune complexes, phagocytosis of IgG-coated particles, and release of inflammatory agents.

Mouse monoclonal anti-DNP IgG1 antibodies were purified from supernatants of the mouse-mouse hybridomas ANO2 (Balakrishnan et al., 1982), 1B7.11 (Kimura et al., 1986) and DHK109.3 (Liu et al., 1980) by affinity chromatography with DNP-HSA (see above) and were conjugated with tetramethylrhodamine isothiocyanate (R-). Solution association constants for DNP-G were measured by ultraviolet fluorescence quenching (Pisarchick and Thompson, 1990) and were not significantly different for labeled and unlabeled antibodies (Table III).

All antibodies were clarified by sedimentation at 100,000 g and 4 °C for 2 hours before use with planar membranes. Gel filtration of mouse IgG and ANO2 indicated that, in this case, both labeled and unlabeled antibodies eluted with symmetrical peaks and at identical volumes corresponding to IgG monomers. IgG solution concentrations were determined after clarification, both spectrophotometrically and by a bicinchoninic acid (photometric) assay; both methods gave equivalent results.

MoFcγRII was purified from J774A.1 homogenates using affinity chromatography with the Fabs of a monoclonal anti-moFcγRII monoclonal antibody (2.4G2; Mellman and Unkeless, 1980). SDS-PAGE, Western blots with 2.4G2 as the primary antibody, and N-terminal amino acid sequencing identified the purified protein as moFcγRII (Poglitsch et al., 1991). Purified moFcγRII was reconstituted into egg phosphatidylcholine/cholesterol vesicles (6:1, w/w), at a protein/lipid ratio of 1:10 (w/w), by detergent dialysis.

Substrate-supported planar membranes containing moFcγRII [moFcγRII(+)] or not containing moFcγRII [moFcγRII(-)] were formed by treating fused quartz surfaces with vesicle suspensions (Poglitsch et al., 1991). Previous work using vesicles doped with fluorescent lipids has demonstrated that this procedure yields a uniformly fluorescent surface coating in which the fluorescent lipids display long range lateral mobility with an apparent diffusion coefficient typical of fluid-like phospholipid bilayers ($\approx 10^{-8}$ cm^2/sec) (Poglitsch et al., 1991).

For microscopy, planar membranes were washed with PBS, treated with bovine serum albumin (BSA) in PBS (10 mg/mL, 30 min), and then treated with R-IgG (concentration as specified, 30 min) in 10 mg/mL BSA/PBS. In some samples, the R-IgG solutions contained

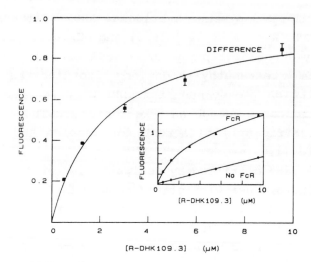

<u>Figure 4</u>. <u>R-DHK109.3 on Planar Membranes</u>. The difference (■) of the evanescently excited fluorescence of R-DHK109.3 on moFcγRII(+) (▲) and moFcγRII(-) (◆) membranes increases in a saturating manner as a function of the solution concentration of R-DHK109.3. Shown also is the best fit of the difference data to Eq. 6 (———). Uncertainties are standard errors in the means for averages over ten spatially independent measurements on each of three planar membranes. Parameters for TIRFM measurements were as follows: illuminated area, approximately 20 μm x 200 μm; laser wavelength, 514.5 nm; laser power, 50 μW; temperature, 25-28 °C.

100 μM DNP-G, which is sufficient to saturate ≥ 95% of the IgG DNP binding sites. Under these conditions, R-IgG specifically binds to moFcγRII in the membranes (in that the absence of moFcγRII or the Fc region of IgG significantly reduces the fluorescence to equivalently low values), the R-IgG surface association reaches a steady-state, and the R-IgG solution concentrations adjacent to the planar membranes are equal to the applied R-IgG concentrations. In addition, cross inhibition measurements show that unlabeled 2.4G2 Fab blocks the binding of R-IgG and that unlabeled IgG (partially) blocks the binding of R-(2.4G2 Fab) (Poglitsch et al., 1991).

The fluorescence measured with TIRFM of anti-DNP R-IgG antibodies on moFcγRII(+) membranes increased as a function of the R-IgG solution concentration, whereas the fluorescence on moFcγRII(-) membranes was significantly lower and linear with concentration (Figure 4). Apparent association constants K were determined by subtracting the fluorescence intensities on

TABLE III: ASSOCIATION CONSTANTS FOR IgG WITH DNP-G AND moFc RII

Antibody	DNP-G in Solution	moFcγRII in Membranes		
		DNP-G(-)	DNP-G(+)	Density
1B7.11	68 ± 25	2.6 ± 1.3	3.0 ± 1.2	680 ± 150
ANO2	29 ± 2	8.1 ± 1.8	3.8 ± 1.6	540 ± 120
DHK109.3	2.5 ± 0.4	3.9 ± 0.5	4.8 ± 1.1	660 ± 50
mouse IgG	-	2.9 ± 0.4	3.1 ± 0.6	430 ± 80

Values in the first column are the means and standard errors in the means of six independent quenching curves; data were corrected for collisional quenching using the average of four curves for mouse IgG. Values in the last three columns are the means and standard errors in the mean of three independently obtained and analyzed binding curves (see Figure 4). The solution concentration of DNP-G was 100μM. Association constants are in units of 10^5 M^{-1} and antibody surface site densities (calculated from data obtained in the absence of DNP-G) are in units of molec/μm^2. All association constants were determined at 25-28°C.

moFcγRII(-) membranes (containing only solution-phase R-IgG) from those on moFcγRII(+) membranes (containing both membrane-bound and solution-phase R-IgG) and fitting the fluorescence differences to the functional form for a reversible, bimolecular reaction (Eq. 6). Antibody surface site densities [N] were estimated from the ratios of the fluorescence intensities for R-IgG on moFcγRII(+) membranes to those on control membranes (Eq. 7).

The measured association constants for IgG with reconstituted FcγRII are summarized in Table III. The association constants for polyclonal IgG and monoclonal IgG1 are similar and agree well with values obtained indirectly for myeloma IgG1 with macrophage cell surfaces using radioimmune assays at 0°C (Segal and Titus), 1978).

The association constants in the presence and absence of saturating amounts of DNP-G are not statistically different for polyclonal IgG, 1B7.11 or DHK109.3, while a small difference exists in the presence and absence of DNP-G for ANO2. These results suggest that, for moFcγRII in planar membranes, the association of DNP-G with anti-DNP IgG1 monoclonal antibodies does not significantly alter the IgG - Fc mediated association with moFcγRII. Hapten-induced conformational changes that do not significantly affect the association constant are not precluded; however, previous work suggests that the occupation of IgG binding sites with small haptens does not significantly perturb Fc region structure or Fc-mediated effector function (Burton, 1985; Colman, 1988). Time-resolved fluorescence polarization studies have indicated that the IgG Fc domain may undergo structural changes with larger ligands (Schlessinger et al., 1975), and further investigations of the effects of monovalent peptide antigens on IgG-moFcγRII interactions may prove to be interesting.

CONCLUSION: Understanding the transport, equilibrium and kinetic behavior of antibodies at cell membrane surfaces is central to a complete physical description of antibody-mediated cellular immune processes. The work described in this paper has shown that a variety of techniques in fluorescence microscopy can be used to quantitatively examine the dynamics of antibody interactions with substrate-supported phospholipid membranes that mimic cell surfaces to which antibodies bind through their antigen binding sites or through their Fc regions.

Acknowledgements: We thank Alan Waggoner of Carnegie-Mellon University for CY3.18; Harden McConnell of Stanford University for ANO2 hybridomas; Betty Diamond of the Albert Einstein College of Medicine for 2.4G2 hybridomas; and Norman Klinman of the Scripps Research Institute for DHK109.3 hybridomas. J774A.1 and 1B7.11 cells were obtained from the American Type Culture Collection. Support was provided by NIH Grant GM-37145 and by a Camille and Henry Dreyfus Teacher-Scholar Award.

REFERENCES

Axelrod, D., Koppel, D. E., Schlessinger, J., Elson, E. and Webb, W. W. (1976) Biophys. J. 16, 1055-1069.
Axelrod, D., Burghardt, T. P., and Thompson, N. L. (1984) Ann. Rev. Biophys. Bioeng. 13, 247-268.
Balakrishnan, K. F., Hsu, F. J., Cooper, A. D., and McConnell, H. M. (1982) J. Biol. Chem. 257, 6427-6433.

Burghardt, T. P. and Axelrod, D. (1981) Biophys. J. 33, 455-468.
Burton, D. R. (1985) Mol. Immunol. 22, 161-206.
Colman, P. M. (1988) Adv. Immunol. 43, 99-132.
Ethier, M. F., Wolf, D. E., and Melchior, D. L. (1983)
    Biochemistry 22, 1178-1182.
Herron, J. N., Kranz, D. M., Jameson, D. M. and Voss, E. W. (1986)
    Biochemistry 25, 4602-4609.
Johnson, P. and Garland, P. B. (1983) FEBS Lett. 153, 391-394.
Kimura, K., Nakanishi, M., Ueda, M. Ueno, J., Nariuchi, H.,
    Furukawa, S., Yasuda, T. (1986) Immunology 59, 235-238.
Kinosita, K., Kawato, S., and Ikegami, A. (1977) Biophys. J. 20,
    289-305.
Liu, F., Bohn, J. W., Ferry, E. L., Yamamoto, H., Molinaro, C. A.,
    Sherman, L. A. Klinman, N. R. and Katz, D. H. (1980) J. Immunol.
    124, 2728-2737.
McConnell, H. M., Tamm, L. K. and Weis, R. M. (1984) Proc. Natl.
    Acad. Sci. U.S.A. 81, 3249-3253.
McConnell, H. M., Watts, T. H., Weis, R. M. and Brian, A. A.
    (1986) Biochim. Biophys. Acta 864, 95-106.
Mellman, I. S., and Unkeless, J. C. (1980) J. Exp. Med. 152, 1048-
    1069.
Mellman, I., Koch, T., Healey, G., Hunziker, W., Lewis, V.,
    Plutner, H., Miettinen, H., Vaux, D., Moore, K., and Stuart, S.
    (1988) J. Cell Sci. Suppl. 9, 45-65.
Pisarchick, M. L. and Thompson, N. L. (1990) Biophys. J. 58, 1235-
    1249.
Poglitsch, C. L., and Thompson, N. L. (1990) Biochemistry 29, 248-
    254.
Poglitsch, C. L., Sumner, M. T. and Thompson, N. L. (1991)
    Biochemistry, in press.
Schlessinger, J., Steinberg, I. Z., Givol, D., Hochman, J. and
    Pecht, I. (1975) Proc. Natl. Acad. Sci. U.S.A. 72, 2775-2779.
Segal, D. M. and J. A. Titus (1978) J. Immunol. 120, 1395-1403.
Smith, B. A. and McConnell, H. M. (1978) Proc. Natl. Acad. Sci.
    U.S.A. 75, 2759-2763.
Smith, L. M., Weis, R. M. and McConnell, H. M. (1981) Biophys. J.
    36, 73-91.
Theriault, T. P., Rule, G. S., and McConnell, H. M. (1991)
    Biophys. J. 59, 189a.
Thompson, N. L., Burghardt, T. P. and Axelrod, D. (1981) Biophys.
    J. 33, 435-454.
Thompson, N. L., Palmer, A. G. Wright, L. L. and Scarborough, P.
    E. (1988) Comm. Molec. Cell. Biophys. 5, 109-131.
Timbs, M. M., Poglitsch, C. L., Pisarchick, M. L., Sumner, M. T.
    and Thompson, N. L. (1991) Biochim. Biophys. Acta, in press.
Timbs, M. M., and Thompson, N. L. (1990) Biophys. J. 58, 413-428.
Unkeless, J. C., Scigliano, E., and Freedman, V. H. (1988) Ann.
    Rev. Immunol. 6, 251-281.
Uzgiris, E. E. and Kornberg, R. D. (1983) Nature 301, 125-129.
Velez, M. and Axelrod, D. (1988) Biophys. J. 53, 575-591.
Wagner, M., Ernst, L., Majumdar, R., Majumdar, S., Chao, J. and
    Waggoner, A. (1990) 14th International Meeting of the Society
    for Analytical Cytology, Asheville, NC, Abstract #459.
Wright, L. L., Palmer, A. G. and Thompson, N. L. (1988) Biophys.
    J. 54, 463-470.

Progress in Membrane Biotechnology
Gomez-Fernandez/Chapman/Packer (eds.)
© 1991 Birkhäuser Verlag Basel/Switzerland

LOCATION AND DYNAMICS OF α-TOCOPHEROL IN MEMBRANES

Juan C. Gómez-Fernández, Francisco J. Aranda and José Villalaín

Department of Biochemistry and Molecular Biology, University of Murcia, E-30071 Murcia, Spain.

SUMMARY

α-Tocopherol significantly perturbs the phase transition of fully saturated phospholipids and its effect is considerably bigger than that of α-tocopheryl acetate as observed by differential scanning calorimetry and Fourier-transform infrared spectroscopy. This observation remarks the importance of its phenolic hydroxyl group in determining the interaction of α-tocopherol with phospholipids. However, this hydroxyl group is not hydrogen bonded to neither the carbonyl group nor the phosphate group of phosphatidylcholine, although it seems able of forming a hydrogen bonding with the carbonyl group of phosphatidylethanolamine. By studying the intrinsic fluorescence of α-tocopherol, it was found that its chromanol group is located near the water-lipid interface, although it is not accesible from the water phase. On the other hand, α-tocopherol was found to partition with preference for the most fluid domains in the membrane, as seen by differential scanning calorimetry.
    The dynamics of α-tocopherol in membranes was examined through the quenching of its intrinsic fluorescence and it was found to have a diffusion coefficient similar to that of phospholipids.
    α-Tocopherol was also found to promote the H$_{II}$ phase of phosphatidylethanolamine, having a stronger effect than α-tocopheryl acetate.
    Finally, α-tocopherol was shown to have a stabilizing effect on liposomes, when they were stored for periods of time longer than one year.

INTRODUCTION

α-Tocopherol (α-T) is a very important component of biological membranes where it act as a stabilizer (Tappel, 1972). The main stabilization effect seems to be the inhibition of chain peroxidation of membrane lipids (Scott, 1978; DeDuve and Hayaishi, 1978).

In order to understand the molecular mechanism of action of α-T we have undertaken over the past years a systematic study of the interaction of α-T with phospholipids when incorporated in model membranes, and using a number of physical techniques as it is shown below.

MODULATION OF MEMBRANE FLUIDITY BY TOCOPHEROLS

Tocopherols are able of modulating the phase transitions of phospholipids, as appreciated by differential scanning calorimetry. The effect of α-tocopherol (α-T) on DPPC was found to be bigger than that of α-tocopheryl acetate (α-TA) (Fig. 1), so that the hydroxyl group seems to be very important for the interaction of tocopherols with phospholipids. The main phase transition of saturated phosphatidylcholines is broadened and shifted to lower temperatures in the presence of α-T. The shift

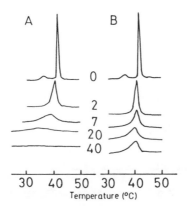

Fig. 1. The DSC calorimetric curves for systems containing pure DPPC and (a) α-tocopherol and (b) α-tocopheryl acetate. Molar percentages of vitamin E in DPPC are indicated on the curves.

to lower temperatures of Tc indicates that α-T will prefer to partition into the most fluid domains, which is also confirmed by other studies (see below).

The effect of tocopherols on the phase transition of DPPC was also examined using Fourier transform infrared spectroscopy (FT-IR). Infrared spectroscopy is a very well suited technique to characterize the degree of fluidity in membranes (Casal and Mantsch, 1984). The frequency of the CH_2 antisymmetric stretching vibration mode of the acyl chains of the phospholipid, when plotted versus temperature allows the observation of the phase transition from all-*trans* to *gauche* conformers (Casal and Mantsch, 1984) and hence an estimation of the average number of *gauche* conformers. Fig. 2 shows that the phase transition of DPPC is remarkably altered, as seen through the frequency of this band, in the presence of α-T and α-TA. Increasing concentrations of α-T produce a progressive broadening of the phase transition and a shift of the onset of this transition from $41°C$ in pure DPPC to $34°C$. At temperatures above the phase transition there is a progressive decrease in *gauche* isomers as the concentration of α-T was increased, this effect being the same type as the effect given by cholesterol (Cortijo et al., 1982). However, α-T behaves differently from cholesterol when temperatures below the phase transition are considered, since the increase in *gauche* isomers produced by cholesterol is not observed here, but instead a

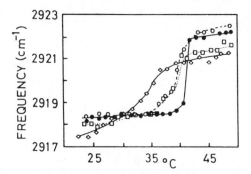

FIG. 2. Frequency of the CH2 antisymmetric stretching mode of DPPC vs temperature. (●━━━●), Pure DPPC, (O------O) 20 mol% of α-tocopheryl acetate, (□··············□) 5 mol% of α-tocopherol and (◇━━━◇) 20 mol% of α-tocopherol in DPPC.

decrease in frequency was found which might indicate a decrease in the percentage of *gauche* isomers. Perhaps the very close interaction established between α-T and the phospholipid may explain why the acyl chains are further immobilized with respect to the gel phase of the pure phospholipid.

The incorporation of increasing concentrations of α-T produces a progressive broadening of the phase transition and a shift of the onset of this transition to lower temperatures. Both above and below the phospholipid phase transition there is a decrease in *gauche* isomers. α-TA broadens the phase transition, however, but does not appreciably affect the proportion of *gauche* conformers at the very high concentration tested of 20 mole percent (Villalaín et al., 1986).

LOCATION OF α-TOCOPHEROL IN PHOSPHOLIPID VESICLES

It is very important to study the location of α-T in membranes in order to better understand the molecular mechanism of action of α-T as a membrane stabilizing agent. We have tried to provide evidence on the location of α-T in membranes using FT-IR by monitoring its interaction with phospholipids, specifically DPPC and DMPE.

It has been suggested that α-T is situated in the membrane with the phenolic hydroxyl group near the polar moiety of the lipid matrix (Perly et al., 1985) and a polar interaction between α-T and phospholipids has also been proposed (Baig and Laidman, 1983). It has also been suggested the existence of a hydrogen bond interaction between the hydroxyl group of α-T and the polar region of phospholipids (Srivastava et al., 1983). We have addressed this question by studying the interaction of α-T and α-TA with DPPC and DMPE in order to see, first, if there are any differences on the effect of both tocopherols and second, if these differences can be adscribed to the existence of a specific interaction between tocopherols and the phospholipids.

The perturbation of the membrane interfacial region can be followed by monitoring the frequency of the maximum of the C=O stretching band and the frequency of its components resolved by enhancement methods (Casal and Mantsch, 1984). We have found that α-T induces a decrease in the frequency of the maximum of this band at all temperatures and this effect is more pronounced than that produced by α-TA. In order to see if this change in frequency of the hydrated components upon α-T inclusion is due to a specific interaction, we have used DPPC which is specifically substituted with ^{13}C at the sn-2 C=O chain group ((sn-2-1-^{13}C)-DPPC). The substitution of the sn-2 group with ^{13}C allows its independent observation from the sn-1 group since its stretching band is shifted 45 cm^{-1} to lower frequency, due to the vibrational isotope effect, with respect to the band corresponding to the sn-1 carbonyl group (Blume et al., 1988). We have used this approach to study the interaction of α-T and α-TA with DPPC as shown in Fig. 3. The inclusion of α-T or α-TA does not induce any appreciable change in the frequencies of the component bands, neither on the band contour of the sn-1 and sn-2 bands nor on the hydrated and dehydrated components.

As a comparison of the effect of α-T in DPPC the effects of cholesterol on DPPC are shown in Fig. 4. Cholesterol is a biomolecule similar to α-T in the sense that both of them have a free hydroxyl group which could be implicated in hydrogen bonding. As we can see in Fig. 4, where the C=O group region of

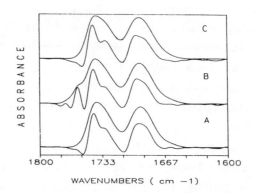

Fig. 3. Infrared spectra of the C=O stretching region of (sn-2-1-^{13}C)-DPPC. (A) Pure (sn-2-1-^{13}C)-DPPC, (B) (sn-2-1-^{13}C)-DPPC plus 10 mol% of α-tocopheryl acetate and (C) (sn-2-1-^{13}C)-DPPC plus 10 mol% of α-tocopherol.

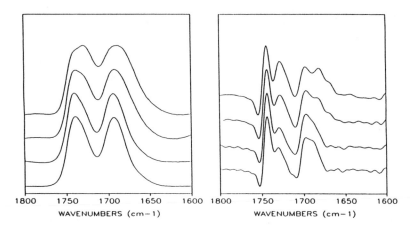

1800   1750   1700   1650   1600   1800   1750   1700   1650   1600
WAVENUMBERS (cm−1)   WAVENUMBERS (cm−1)

Fig. 4. Infrared spectra of the C=O stretching region of (sn-2-1-^{13}C)-DPPC. From top to bottom, (sn-2-1-^{13}C)-DPPC:cholesterol 1:1, (sn-2-1-^{13}C)-DPPC:cholesterol 5:1, (sn-2-1-^{13}C)-DPPC:cholesterol 8:1 and pure (sn-2-1-^{13}C)-DPPC. Left and right panels present the spectra and the deconvoluted spectra respectively.

(sn-2-1-^{13}C)-DPPC is shown, the inclusion of increasing concentrations of cholesterol leads to a change in the sn-2 band but not in the sn-1. This change can be described as a shift from the band at 1687 cm^{-1} in pure (sn-2-1-^{13}C)-DPPC to 1679 cm^{-1} in the presence of cholesterol 1:1 to (sn-2-1-^{13}C)-DPPC and the appearance of a new band at lower frequencies, at 1658 cm^{-1}. The band at 1679 cm^{-1} is suggested to be due to a sn-2 carbonyl group hydrogen bonded to cholesterol and the 1658 cm^{-1} component to sn-2 carbonyl groups hydrogen bonded to both a water molecules and cholesterol. From a comparison of Fig. 3 and Fig. 4, it can be concluded that, whereas cholesterol interacts specifically with the sn-2 carbonyl group of DPPC through the formation of a hydrogen bond, α-T does not establish any specific bonding with the carbonyls of DPPC.

The PO_2^- antisymmetric stretching band of DPPC which appears at 1220 cm^{-1} was also observed in the presence of α-T and α-TA, in order to check whether tocopherols could establish hydrogen bonding with this group. It was found that this band was not affected at all neither by the presence of α-T nor α-TA (results not shown). It seems then that tocopherols do not establish any hydrogen bonding with DPPC.

The interaction of α-T and α-TA with DMPE was also studied and it was found that both types of molecules perturb the intermolecular hydrogen bond network which has been described to be present in DMPE. Hydrated DMPE presents a broad C=O stretching band which after derivation can be resolved in two components, one at 1744 cm^{-1} and the other at 1720 cm^{-1} (Fig 5).

The component at 1744 cm^{-1} can be assigned to a dehydrated C=O group, whereas the component at 1720 cm^{-1} can be assigned to a hydrogen bonded C=O group (Wong and Mantsch, 1988). The inclusion of α-T and α-TA lead to significant changes in the band contour of the C=O band of DMPE, as shown in Fig. 6 A and B.

In the presence of α-T and after resolution enhancement methods, three main components are apparent, whose frequencies are 1742, 1734 and 1720 cm^{-1} (Fig. 6A). However, the inclusion of α-TA produces a different pattern, only two components at 1744 and 1734 cm^{-1} are seen, although a broad component at lower frequencies in the original spectra is barely discernible (Fig. 6B). The interpretation of these results is that the hydrogen bond network which is present in pure DMPE is also observed in DMPE plus α-T, so that α-T may also participate in this hydrogen bonding with C=O groups of DMPE. α-TA, however, seems to disrupt the hydrogen bonding interactions occurring in pure DMPE, perhaps simply by its interposition between DMPE molecules. Another group

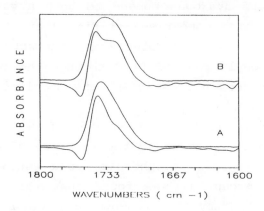

Fig. 5. Infrared spectra of the C=O stretching region of DMPE.

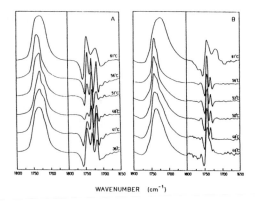

Fig. 6. Infrared spectra of the C=O stretching region of DMPE. (A) 10 mol% of α-tocopherol and (B) 10 mol% of α-tocopheryl acetate in DMPE.

of DMPE that could be involved in hydrogen bonding with tocopherols is the phosphate group. We have not observed any significant differences in the PO_2^- antisymmetric stretching band of DMPE neither in the presence of α-T nor of α-TA, but significantly, and as seen in Fig. 7, where the PO_2^- antisymmetric stretching region of DMPE is shown, the inclusion of α-T induces a reduction in the bandwidth of this band, which is not observed when α-TA is included in DMPE, demonstrating very clearly that α-T induces a rigidification of the PO_2^- group of DMPE, although this rigidification is not a consequence of a specific interaction of α-T with the PO_2^- group of DMPE.

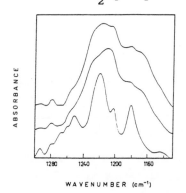

Fig. 7. Infrared spectra of the PO_2^- antisymmetric stretching vibration band of DMPE. (A) Pure DMPE, (B) DMPE plus 10% α-tocopheryl acetate and (C) DMPE plus 10 mol% α-tocopherol.

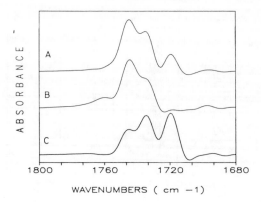

Fig. 8. Infrared spectra of the C=O stretching region of anhydrous DMPE. (A) Pure DMPE, (B) 10 mol% of α-tocopherol and (C) 10 mol% of α-tocopheryl acetate in DMPE.

The effect of α-T and α-TA on DMPE has also been studied in anhydrous mixtures and in Fig. 8 it is shown the C=O stretching vibration region of anhydrous DMPE. In this state, pure DMPE presents three main bands at 1745, 1735 and 1720 cm^{-1}, where the 1720 cm^{-1} band is due to an hydrogen bonded carbonyl group. In the presence of α-T the area of this band is increased, but it is absent in the presence of α-TA. Furthermore, the PO_2^- antisymmetric stretching band is split in two bands in the presence of α-T but not in the presence of α-TA (Fig. 9), which confirms that α-T disrupts the intermolecular hydrogen bonds of DMPE and, probably, forms itself hydrogen bonds with the phospholipid.

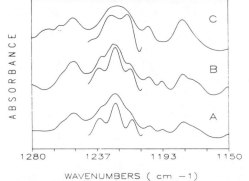

Fig. 9. Infrared spectra of the PO_2^- antisymmetric stretching vibration band of anhydrous DMPE. (A) Pure DMPE, (B) DMPE plus 10% α-tocopheryl acetate and (C) DMPE plus 10 mol% α-tocopherol.

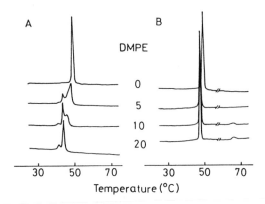

Fig. 10. The DSC calorimetric curves for mixtures of (A)
DMPE/α-tocopherol and (B) DMPE/α-tocopheryl acetate. Molar
percentages of vitamin E are indicated on the curves.

The situation of the hydroxyl group of α-T in the
lipid/water interface is of interest in explaining its mechanism
of action because thanks to this location it could prevent the
introduction in the membrane of oxidizing agents.

The effect of tocopherols on phosphatidylethanolamines has
been also studied by DSC (Micol et al., 1990). α-T induced the
appearance of more than one peak transition for the gel-to-liquid
crystalline in DMPE (Fig. 10A), indicating that immiscibilities
were taking place, so that phases with different contents in α-T
and phospholipid are present, the Tc transition temperature being
lower as more α-T is present in each particular phase. However,
α-TA did not produced this effect, and only one transition peak
was observed for the gel-to-liquid crystalline phase transition
(Fig. 10B).

Another important factor to be understood is whether α-T is
randomly distributed in the membrane or if it shows a preference
for certain regions or domains. It is pertinent to this question
to remember that it has been already commented above, that α-T
shows a preference for the most fluid domains in the membrane.

This preference was also confirmed by other DSC studies
where α-T was included in mixtures of phospholipids showing
monotectic behavior such as DMPC/DPPE and DLPE/DSPC (Fig 11). In
all these systems α-T always affected most to the component

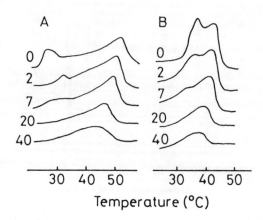

Fig. 11. The DSC calorimetric curves for mixtures of (a) DMPC/DPPE and (b) DLPE/DSPC containing different molar % of α-tocopherol as indicated.

having the lowest phase transition T_c, irrespectively of whether this was PC or PE (Ortiz et al., 1987). Since it seems very likely that biological membranes are composed of dominions having different degrees of fluidity it is very important to keep in mind that α-T will be preferentially found in those which are the most fluid. Furthermore, this may have important implications with respect to the mechanism of action of α-T. That is, unsaturated fatty acyl chains are to be predominantly found in the most fluid domains, and at the same time they are the lipid component prone to oxidation. Since α-T will prevent chain peroxidations, the regions of the membrane where its presence will be meaningful will be precisely these fluid domains. Hence thinking from the teleological point of view, α-T is correctly designed since it will be accumulated in the place where it could be most useful.

DYNAMICS OF α-TOCOPHEROL IN MEMBRANES

We have used the intrinsic fluorescence of α-T to determine some basic fluorescence parameters of α-T in a number of solvents and incorporated into phospholipid vesicles (Aranda et al., 1989). We have found that both quantum yield and fluorescence lifetime of α-T increase as solvent polarity increases. It is interesting to note that λ_{max} and fluorescence lifetime obtained for α-T in phospholipid vesicles (316 nm and 1.7 ns) are very similar to those obtained in protic solvents like ethanol. This observation suggest that the chromanol moiety of α-T should be located in the polar region of the model membrane.

The absorption spectra of α-T in n-hexane show a maximum at 283 nm for low concentration of α-T (predominance of the monomeric form), but it is shifted to 295 nm at much high concentration (predominance of the hydrogen bond). α-T in phospholipid vesicles showed a maximum near 295 nm within a wide range of concentration, indicating that most α-T molecules are associated when present in membranes. However, we have found that although α-T molecules are associated, the aggregates formed are fluorescent, since a linear relationship dependence between fluorescence intensity and a large range of concentrations is obtained both in homogeneous media and in membranes.

The location of α-T in the bilayer was also studied through the quenching of its fluorescence by membrane probes such as 5-NS and 16-NS. We have found that 5-NS quenches α-T fluorescence much more effectively than 16-NS, as should be expected if the chromanol moiety is located near the lipid/water interface.

We have also studied the location of α-T in phospholipid vesicles by using a set of n-(9-anthroyloxy)-stearic acid (n-AS) probes. The explicit distance dependence (r^{-6}) of electronic energy transfer (dipolar mechanism) has allowed its application as a spectroscopic ruler for determining distance in biological systems. Figure 12 shows the efficiency of energy transfer from α-T to different n-AS probes differing in the location of their 9-anthroyloxy group. It can be seen that the efficiency of energy transfer follows the order 7-AS > 2-AS > 9-AS = 12-AS. From these data, it can be concluded that the chromophore group of α-T is situated in the membrane in a region between the 9-anthroyloxy group located at carbon 7 and carbon 2, the former being the nearest one. Furthermore, we found that acrylamide, which is a water soluble fluorescence quencher, has a very low efficiency of

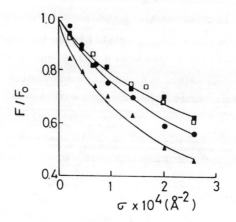

Fig. 12. Relative yield of α-tocopherol fluorescence F/F₀ vs. σ (acceptor surface concentration) of 2-AS (●), 7-AS (▲), 9-AS (■) and 12-AS (□) in egg yolk PC at 25°C. Excitation and emission wavelengths were 295 and 329 nm, respectively.

quenching α-T in phospholipid vesicles, while acrylamide is an efficient quencher of α-T in ethanolic solution (Fig. 13). This indicate that although α-T may have its chromanol group relatively close to the polar part of the bilayer, it is not sufficiently exposed to allow acrylamide to reach it (acrylamide being known to have a very low capacity of penetration through phospholipid bilayers).

The conclusions of these studies is that the chromanol group is found in a position close to that occupied by 7-AS and 5-NS.

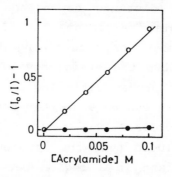

Fig. 13. Stern-Volmer plots of quenching of α-tocopherol fluorescence by acrylamide. in ethanol (○), α-tocopherol concentration 10 μM; and in DPPC at 50° (●), lipid concentration 500 μM and DPPC/α-tocopherol molar ratio 500:1.

The location of the chromanol moiety in the lipid/water interface could be of importance in explaining its mechanism of protection since any oxidizing agent approaching the membrane surface should find reducing protons and its penetration should be avoided.

In order to understand the mechanism of action of α-T in membranes it may be very illustrative to know the lateral diffusion of this molecule when incorporated into phospholipid vesicles. This has been approached through the studies of the quenching of the intrinsic fluorescence of α-T by 5-NS. By using the Smoluchwoski equation we found a lateral diffusion coefficient of 4.8 x 10^{-6} cm^2s^{-1} (Aranda et al., 1989). This means that α-T may have high mobility in natural membranes and hence be quite efficient in reacting with oxidizing agents. This value is very similar to the one calculated by others authors for ubiquinone-3 of 5.8 x 10^{-6} cm^2s^{-1} (Fato et al., 1986) which is a molecule very related in structure to α-T. It is interesting to note that the lateral diffusion coefficient obtained for NBD-phosphatidylcholine using this method as well (unpublished data), is quite similar to that of α-T.

MODULATION BY TOCOPHEROLS OF THE LIPID POLIMORPHIC PROPERTIES OF PHOSPHATIDYLETHANOLAMINES

It is known that phospholipids of membranes, either of biological origin or artificial, can adopt several structures, such as the micellar phase, the bilayer and hexagonal H_{II} phases and lipidic particles. The ability of lipids to adopt these different structures is known as lipid polymorphism (Cullis et al., 1985).

The non-bilayer structures can greatly affect the functional behavior of the membrane (De Kruijff et al., 1985) with a big biological potential significance. We have studied the effect of α-T and α-TA on lipid polymorphism of phosphatidylethanolamine, a phospholipid which can spontaneously adopt the hexagonal H_{II} phase at physiological temperatures. We have used DEPE and DMPE, an unsaturated and a saturated PE. The bilayer to H_{II} transition

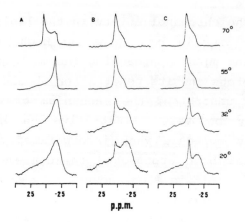

Fig. 14. ^{31}P-NMR spectra of (A) pure DEPE, (B) DEPE containing 10 mol% of α-tocopherol and (C) DEPE containing 10 mol% of α-tocopheryl acetate at different temperatures as indicated.

of DEPE is well characterized occurring at 63 °C, but this transition takes place at much higher temperatures in the case of DMPE (Micol et al., 1990).

Both α-T and α-TA have almost no effect on the gel-to-liquid crystal phase transition of DEPE, but the effect on the bilayer to H_{II} phase transition is more significant, because increasing amounts of both molecules decrease this transition temperature (Micol et al., 1990). ^{31}P-NMR confirms this effect as seen in Fig. 14, which shows that, even below the gel-to-liquid crystal phase transition, a fraction of the DEPE molecules organize themselves in H_{II} structures.

The effect of α-T on DMPE is different from that given on DEPE, whereas the effect of α-TA is similar on both phospholipids. α-Tocopheryl acetate does not affect significantly the gel-to-liquid crystal transition of DMPE, but α-T produces the presence of several transition peaks (fig. 5). This effect is similar to what has been described previously for the interaction of α-T and other PE's (Ortiz et al., 1987). By ^{31}P-NMR we can observe that α-T induces the presence of H_{II} phase coexisting with bilayer structures, but the presence of α-TA produces a H_{II} phase only well above the gel-to-liquid crystal transition. It can also be seen the existence of isotropic signals in the NMR spectra. These isotropic signals can arise from the existence of intermediate states between the bilayer and H_{II} phases as it has been previously noted (Cullis and Kruijff, 1978).

These results demonstrate that tocopherols induce H_{II} phase formation in PE systems, α-T being more effective than α-T acetate. The formation of H_{II} hexagonal phase would be compatible with the increasing motion on the acyl chains induced by tocopherols as shown by FT-IR, which would increase the hydrophobic volume of the lipid, facilitating the formation of the hexagonal structures. The hydroxyl group of α-T would facilitate also its positioning in the bilayer so that the interaction between α-T and the hydrophobic part of the lipid would be maximized, explaining the differences found between α-T and α-T acetate, which has its hydroxyl group blocked by the acetate. It remains to be seen if the influence of tocopherols on lipid polymorphism is significant from the point of view of their biological effects.

STABILITY OF LIPOSOMES ON LONG TERM STORAGE : THE EFFECT OF α-TOCOPHEROL.

Liposome research is an important part in biological, pharmaceutical and medical research because liposomes are one of the most effective carriers for the introduction of many kinds of biomolecules into cells and body fluids. The entrapment of a drug changes its pharmacokinetics and it can result in a better therapeutic index, and the cellular uptake can be enhanced. One of the most important difficulties encountered when liposomes are to be used as a pharmaceutical dosage form is their stability on long term storage.

We have addressed this question and found that α-T stabilizes liposomes and make them much more stable to leakage of encapsulated products (Hernández-Caselles et al., 1990). As seen on Fig. 15, drug retention is dramatically increased when α-T is incorporated into the membrane at all conditions tested. α-Tocopherol can stabilize the phospholipid bilayer through

114

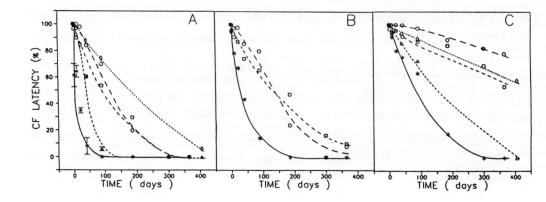

Fig. 15. 5,6-Carboxyfluorescein retention in multilamellar liposomes stored at (A) room temperature, (B) room temperature and O₂-free atmosphere and (C) 4°C. Egg yolk lecithin (✳———✳), egg yolk lecithin:cholesterol 5:1 (Δ---------Δ), egg yolk lecithin:α-tocopherol 1:0.05 (□----------□), egg yolk lecithin:α-tocopherol 1:0.2 (O----------O) and egg yolk lecithin:cholesterol:α-tocopherol 5:1:0.3 (◇·········◇).

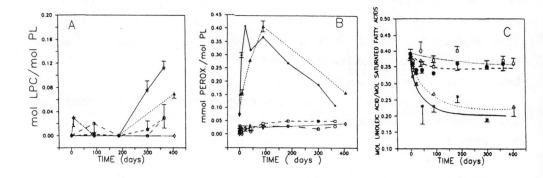

Fig. 16. (A) Lysophosphatidylcholine, (B) Peroxide and (C) Linoleic acid compositional changes in liposomes stored at 4°C. Egg yolk lecithin (✳———✳), egg yolk lecithin:cholesterol 5:1 (Δ----Δ) lecithin:α-tocopherol 1:0.2 (□----□) and egg yolk lecithin:cholesterol:α-tocopherol 5:1:0.3 (◇··········◇).

to ascertain if the compositional changes can be related to leakage of liposome-encapsulated compounds we have measured lysophosphatidylcholine production, peroxide formation and fatty acid degradation as shown in Fig. 16. We have found that lysophosphatidylcholine production (i.e., phospholipid breakdown) in liposomes containing α-T was decreased relative to liposomes without α-T (Fig. 16A). Moreover, peroxide content in liposomes with α-T was minimal at all times and all conditions (Fig 16B) by inhibition of fatty acid peroxidation which would lead to products of degradation which could disrupt the phospholipid bilayer structure leading to leakage of encapsulated products. This is also reflected in the conservation of polyunsaturated fatty acyl chains in the presence of α-T (Fig. 16C).

The effect produced by α-T may be combined with that of cholesterol to enhance the stability of the bilayer (Fig. 15), and hence to improve the storing possibilities of liposomes which encapsulate products.

Acknowledgements

The work described here was supported in part by Grants PA86-0211 and PM90-044 from DGICYT, Spain.

REFERENCES

Aranda, F.J., Coutinho, A., Berberan-Santos, M.N., Prieto, M.J.E. and Gómez-Fernández, J.C. (1989) Biochim. Biophys. Acta 985, 26-32

Baig, M.M.A. and Laidman, D.L. (1983) Biochem. Soc. Trans. 11, 600-602

Blume, A., Hübner, W. and Messner, G. (1988) Biochemistry 27, 8239-8249

Casal, H.L. and Mantsch, H.H. (1984) Biochim. Biophys. Acta 779, 381-401

Cortijo, M., Alonso, A., Gómez-Fernández, J.C. and Chapman, D. (1980) J. Mol. Biol. 157, 597-618

Cullis, P.R. and De Kruijff, B. (1978) Biochim. Biophys. Acta 507, 207-218

Cullis, P.R., Hope, M.J., De Kruijff, B., Verkleij, A.J. and Tilcock, C.P.S. (1985) in Phospholipid and Cellular Regulation (Kuo, J.F., ed.), vol. I, pp. 1-60, CRC Press, Boca Ratón, CA.

DeDuve, C. and Hayaishi, O., eds. (1978), Tocopherol, Oxygen and Biomembranes, Elsevier/North-Holland Medical Press, Amsterdam.

Fato, R., Battino, M., Degli-Esposti, M., Parenti-Castelli, G. and Lenaz, G. (1986) Biochemistry 25, 3376-3390

Hernández-Caselles, T., Villalaín, J. and Gómez-Fernández. J.C. (1990) J. Pharm. Pharmacol. 42, 397-400

De Kruijff, B., Cullis, P.R., Verkleij, A.J., Hope, M.J., Van Echteld, C.J.A. and Taraschi, T.F. (1985) in The Enzymes of Biological Fluids (2nd edn.), (Martonosi, A.N., ed.), vol. I, pp. 131-204, Plenum Press, NY.

Gómez-Fernández, J.C., Villalaín, J., Aranda, F.J., Ortiz, A., Micol, V., Coutinho, A., Berberan-Santos, M.N. and Prieto, M.J.E. (1989) Ann. N. Y. Acad. Sci., USA 570, 109-120

Micol, V., Aranda, F.J., Villalaín, J. and Gómez-Fernández, J.C. (1990) Biochim. Biophys. Acta 1022, 194-202

Ortiz, A., Aranda, F.J. and Gómez-Fernández, J.C. (1987) Biochim. Biophys. Acta 898, 214-222

Perly, B., Smith, I.C.P., Hughes, L., Burton, G.W. and Ingold, K.W. (1985) Biochim. Biophys. Acta 819, 131–135

Scott, M.L. (1978) in The Fat-Soluble Vitamins (DeLuca, H.F., ed.), pp. 133–230, Plenum Press, New York.

Srivastava, S., Phadka, R.S., Goril, G. and Rao, C.N.R. (1983) Biochim. Biophys. Acta 734, 353–362

Tappel, A.L. (1978) Ann. N. Y. Acad. Sci. 205, 12–28.

Villalaín, J., Aranda, F.J. and Gómez-Fernández, J.C. (1986) Eur. J. Biochem. 158, 141–147

Wong, P.T.T. and Mantsch, H.H. (1988) Chem. Phys. Lipids 46, 213–224

Progress in Membrane Biotechnology
Gomez-Fernandez/Chapman/Packer (eds.)
© 1991 Birkhäuser Verlag Basel/Switzerland

PLATELET ACTIVATING FACTOR (PAF), A LIPIDIC MEDIATOR WITH A STRUCTURAL ACTIVITY IN BILAYER.

A MODEL SUSTAINING A BIOLOGICAL RELEVANCE FOR THE ASYMMETRIC PHOSPHOLIPIDS.

Wolf, C., Quinn, P.J., and Chachaty, C.

URA CNRS 1283, CHU Saint Antoine, 27 rue Chaligny, Paris 75012, France.
Biochemistry, King's College London, Campden Hill Road, London W8 7AH, UK.
Physique Générale DPHG/SCM/BP 121, CEN Saclay, 91191 Gif/Yvette, France.

SUMMARY: The structural effect of the lipid mediator PAF is compared with the effect of non-bioactive lysoderivates (lysoPC, lysoPAF). A study using ESR, 31P NMR and X ray diffraction concludes that these asymmetric phospholipids share the same structural activity on bilayers. All of them promote the occurence of interdigitated lamellae and a perturbation of the local director of the lamellar phase. The results agree with the emergence of thin-walled "blisters" within the host phospholipid bilayer after the segregation of the asymmetric phospholipid.

Time-resolved X ray diffraction after a temperature jump to an intermediary temperature between Tc of PAF and Tc of the host lipid (5 to 21°C) indicates that the life-time of the perturbation induced by the asymmetric phospholipid is restricted as a function of the possibility of segregation. The segregation of asymmetric lipids is also restricted by the presence of a low diglyceride fraction (5%). The perturbation brought about by asymmetric lipids is counteracted in the presence of cholesterol. These data are discussed as regard to a structurally active role of the asymmetric phospholipids during the membrane cell activation (played by a lipid mediator like PAF or by products of the phospholipase A activity), which is necessarily restricted in time and in space.

Data obtained after analysis of the molecular species by mass spectrometry of the diglycerides secreted by activated rat platelets support this possibility because it displays a particular enrichment in short acyl chain species, a putative byproduct of phospholipase C activity on asymmetric phospholipid precursors.

INTRODUCTION:

Bioactive lipids, such as PAF (1-O-alkyl,2-acetyl-sn3-glycerophosphorylcholine), prostaglandins, leucotrienes, ... are known to trigger cell activation via a membrane protein receptor (Presscott, S.M. et al, Peplow, P.V. et al.). The PAF receptor has been cloned (Honda, Z.I. et al.), expressed in Xenopus oocytes injected with the transcript, and sequenced. It belongs to the superfamily of G-protein-coupled receptors recognized to stimulate protein kinase C and phospholipase C.

In spite of this apparently complete and definitive description of the mechanism of action, one cannot forget that the lipidic nature of PAF requires a clear deviation from the studies of usual water-soluble agonists. For instance, an extensive so-called "nonspecific" binding has impeded for a long time the disclosure of the membrane receptor.

The structural effect of PAF on the membrane bulk lipids has to be questioned since it could interfere with the activity mediated via the protein receptor. The low activating concentration of PAF (for sensitive cells in the range 0.1-1 nM) but acting on only several hundred to several thousand receptors per cell (Hwang, S.B. et al., Vallari, D.S. et al.) is hardly seen to prevent any interaction of the bioactive lipid with the membrane lipids. On the contrary, it is highly probable that a fraction of the lipid mediator interacts with membrane lipids at some step of the activation process. The lateral diffusion in the membrane being responsible for the distribution of PAF as regard to its respective affinity for membrane lipids or for the protein receptor, the modulation of PAF activity by the membrane lipids is expected and documented (Table I).

120

TABLE I:THE MODULATION OF PAF ACTIVITY BY LIPIDIC FACTORS LEAVES A
POSSIBILITY FOR A LIPID MEDIATOR-MEMBRANE LIPIDS-PROTEIN RECEPTOR
INTERACTION.
1.TRANSLOCATION OF EXOGENOUS PAF OR LYSO-PAF THROUGH PLASMA MEMBRANES
(Tokumura, A. et al., 1990, Biochim. Biophys. Acta, 1044, 91-100).
It influences the degradation of PAF in rabbit platelets or in guinea-pig
leukocytes. It is increased by the preaddition of PAF and by the
temperature.
2.TRANSBILAYER MOVEMENT OF PAF ACROSS HUMAN ERYTROCYTE MEMBRANES
It is increased by the loss of plasma membrane asymmetry (Bratton, DL et
al., 1991, Biochim. Biophys. Acta, 1062,24-34).
3.THE DISTRIBUTION OF PAF WITHIN THE NEUTROPHILS IS ALTERED BY MEMBRANE
FUSION/CYCLING (Riches, DWH. et al., 1990, J. Immunol.).
4.INHIBITION OF PAF-INDUCED PLATELET ACTIVATION BY OLEIC ACID (Nunez, D. et
al., 1990, J. Biol. Chem., 265, 18330-38). Cis unsaturated fatty acids
inhibit instantaneously PAF activation of serotonin release, PIPx
hydrolysis, Protein kinase, and aggregation. But the binding of PAF to the
membrane is not blocked.
5.COUNTERACTION OF BIOPHYSICAL EFFECTS OF PAF BY CHOLESTEROL (olivier et
al.).
6.PAF ANTAGONISTS (some of them beeing very dissimilar to PAF: kadsurenone,
BN 52021, L652731) SHARE COMMON AMPHIPHILIC STRUCTURES (Snyder, F., 1990,
Am. J. Physiol., 259, 697-708).
7.THERE IS A STRUCTURAL "WOBBLING" OF BIOACTIVE PAF MOLECULES (Prescott,
SM. et al., 1990, J. Biol. Chem.,265, 17381-84). Molecular species with a
different lenght and unsaturation of the alkyl chain display the same
activity in vivo. 1 alkenyl 2 acetoyl GPEthanolamine is produced in
stimulated neutrophils and heart muscle cells.

Because PAF binding antagonists do not block universally its actions (Ginka, K.G. et al), non-receptor mechanisms are also questionable.

The stereospecific binding to the protein receptor being the initial event in the activation process, one should be intrigued by the relatively loose structural requirements of the PAF molecule for its cellular activity (O'Flaherty, J.T. et al.). The wide spectrum of PAF antagonists supports also the view that the conformation of the bioactive molecule may "wooble" between certain limits, a view that favours an interaction of PAF within the membrane fluid liquid crystal.

Previous biophysical studies of PAF (Olivier, J.L. et al., Sawyer, D.B. et al.) do not allow to assign any specificity to the structural effect on membrane as compared with non-bioactive lysoderivates. Then, to take into account the preceeding divergent views, we hypothesize that a ternary complex PAF-receptor-membrane lipids could be the efficient agonist in activated cell membranes. This relies also on the modeling of the molecular interaction of PAF with its receptor as computerized to figure out the counteraction of various antiPAF, like BN 52021 (Braquet, P.). Following this model, only the polar group and the glycerol backbone are in close contact with the protein, the alkyl chain being intertwisted with the acyl chains of membrane bulk lipids around the receptor.

The present study has given indications that lysoderivates generated by activated enzymes in stimulated cells could cooperate positively (contrarily to diglycerides) with PAF for the membrane structural activity. A synergy between these different byproducts which appear early in the stimulated cell may be suggested.

Finally, the highly asymmetric PAF (with a large alkyl-acetyl chain lenght inequivalence) can be viewed as a active compound affiliated to a family of asymmetric phospholipids which could be constitutive in biomembranes. The biophysical study *in vitro* of such lipids in model membranes has been carried out previously (for lysoderivates see Van Echteld C.J.A., et al., Killian J.A. et al., Wu, W.G. et al. and Rand, R.P. et al.; for asymmetric lipids see Huang, C. et al.(1983, 1984, 1986), Kramp, W. et al.). But their relevance to the biomembranes under natural conditions is not documented definitively. These asymmetric species could cooperate with PAF to favour the emergence of structural perturbations in the biomembrane.

The present analysis of diglycerides secreted by activated rat platelets supports a particular sensitivity to phospholipase C and therefore the enrichment in asymmetric molecular species of these short-lived byproducts. Recent data (Thuren, T. et al.) are also in agreement with a modulation exerted by PAF on the *in vitro* activity of phospholipase A_2. Then the role of asymmetric phospholipids in membranes, constitutive or appearing transiently as a mediator, could be suggested as a regulation factor of enzyme activities.

MATERIALS AND METHODS

Lipids have been obtained from Sigma (USA) with the exception of PAF (1-O-hexadecyl) and lysoPAF provided by Novabiochem (Switzerland). The spinprobes have been obtained from Molecular Probes (Eugene, USA).

ESR experiments: The build-up of regular multilayer films is detailled by Olivier, J.L. et al. (1988, 1989, 1991). The complete and controled hydration of the films over a salt solution in a sealed vial throughout the measurement is highly recommandable. The glass slide supporting the multilayers is alternatively oriented in the magnetic field in two perpendicular directions, the orientation reference being the normal to the bilayer. ESR spectra are recorded at room temperature with a Bruker 200D with modulation 1 Gpp, microwave (band X) 20 mW. The spinprobe is 5 doxyl stearic acid (5NS) at 0.8% relative to total lipid (0.5 micromole).

For nitroxyde spinprobes, the axis of the $2p_z$ orbital of the odd electron is referred to as the z axis, the x axis being nearly coincident with the N-O bond. The x, y, and z axis are the principal axis of the hyperfine coupling and g tensors. For 5NS, the z axis is on the average (as a function of the conformation variability of this flexible molecule) parallel to the molecular long axis D_M of the rotational diffusion tensor (the equivalent ellipsoid to the spinprobe in motion in its anisotropic environment) (Figure 1). The time-averaged orientation of the molecular axis (D_M) with respect to the local director of the phospholipid bilayer (D_L) is defined by the order parameter $S_{33}=1/2<3\cos^2\beta - 1>=<P_2(\cos \beta)>$, where β is the time-dependent angle between the molecular axis (D_M) and the

the local director (D_L). β is assumed to depend on an energy potential of the form $\exp(-\lambda \cdot P_2(\cos \beta))$. The orientation of the lipid domain is defined by the angle θ between the local director and the normal to the glass slide, N. The average value of θ is zero for a smectic liquid crystal A such as ovolecithin over Tc (no tilt of the acyl chains), however the local director (D_L) is not unique and it is distributed symmetrically around this direction (in the present case, N). The distribution of the orientations of the local director is assumed to be Gaussian with a standard deviation $\Delta\theta$. Figure 1 displays the different angles which are available after computerization of the ESR spectra with the programs described by Chachaty (1985, 1991). These programs hold under the fast motional regime, which, for $P_2(\cos \beta)=0.5$, is extended up to 4 ns. This was checked out using a recent version of the simulation program kindly provided by Dr. J. Freed which holds under the slow motional regime (Schneider, D.J. et al., 1989). In the computed linewidths as a function of the motional correlation times, a contribution (1 G) independent on the motion and on mI is added to take into account the modulation broadening and the unresolved hyperfine proton couplings.

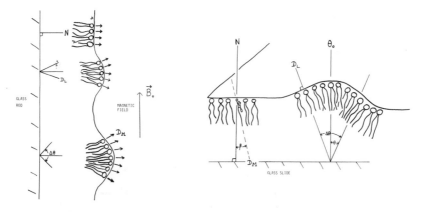

Figure 1: Parameters used to fit ESR spectra by computer simulation. A particular focuss is given on the difference between the (local) order parameter S_{33} qualifying the wobbling of the molecular axis and the disorientation of the local director of the lamella.

Figure 2: Position of the sample (concentric lamellae spread on a glass rod) relative to the magnetic field B in ^{31}P NMR experiments. The field being orthogonal to the averaged molecular axis, we detect the perpendicular component of the averaged ^{31}P Chemical Shielding Tensor. This component is influenced by the curvature of the lamellae where the phospholipids diffuse.

124

NMR Experiments: 400mg mixed lipids in organic solvents are spread over the outer surface of a 5 mm (OD) tube, dried, and hydrated after enclosing the tube into a 10 mm tube over a saturated salt solution. 31P NMR spectra are recorded at 202.7 MHz in a Bruker WM500 spectrometer (vertical sample tube, vertical magnetic field as indicated in Fig. 2). The program used to compute the linewidth of the resonance line of the perpendicular component of the chemical shielding tensor for these oriented lipid films is described by Chachaty (1986).

X ray diffraction study: X ray experiments were carried out using a monochromatic (1.5 A) focussed X ray beam at the Daresbury Synchrotron Laboratory. Lipids hydrated with excess water were sealed between micas 1 mm apart and the cell holder was temperature-controled by a flow of heated gazeous nitrogen. The efficiency of the temperature control and the brightness of the beam allow currently the recording of diffractograms in less than 2.5 seconds with 20 mg lipids during temperature jump experiments with a reaso nable noise/signal ratio. X ray scatterings are processed and plotted as a function of reciprocal spacing ($S=2.\sin\theta/\lambda$) by the Otoko program (Daresbury Laboratories).

Mass Spectrometry: Molecular species of diglycerides secreted by thrombin-(2U/ml; 30 sec)-stimulated rat platelets and of the various platelet phospholipids are obtained by Desorption Chemical Ionization Mass Spectrometry (DCIMS) using ammonia as the reagent gas. Experimental details and the appropriate quantitative control experiments versus HPLC referenced techniques are given in Masrar, H. et al. and follow the method given by Blank et al. for the preparation of the benzoylated derivates.

RESULTS:

Perturbations of the phospholipid bilayer by PAF and other highly asymmetric phospholipids (lysolecithin and lysoPAF):

At concentration far below the micellization of the membrane bilayer, asymmetric phospholipids disturb efficiently the bilayer.

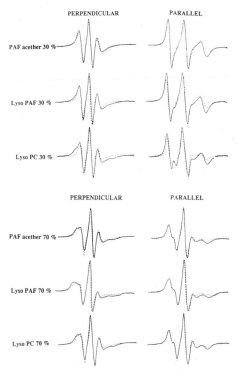

Figure 3: ESR spectra of 5 nitroxide stearic acid probing oriented multibilayers of ovolecithin enriched in asymmetric phospholipids. The spectra are recorded at room temperature under controled hydration. The spectra are simulated (dashed line) according to the parameters indicated in Fig. 1.

Figure 3 displays the alterations of the ESR spectra of 5NS embedded in oriented multibilayers. Alterations are observed from 20% (mole/mole total lipid) for lysolecithin mixed with ovolecithin at room temperature. The spectra on Fig. 3 are presented for 30% and 70% asymmetric lipids, and the computer simulation (dashed line) releases a quantitative view of the spectral alterations in the whole range 0-100% (Figure 4).

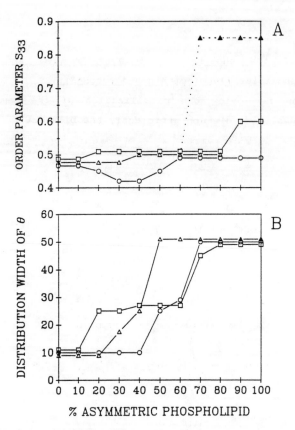

Figure 4: Parameters (order parameter and distribution width of the local director orientation) used to fit ESR spectra of hydrated multibilayers of ovolecithin enriched in PAF (○), lysoPAF (△) or lysoPC (□) probed with 5NS (0.8%). The spectra (band X) are recorded at room temperature, digitalized and computerized after appropriate fitting of the variable parameters depicted in Fig.1.

One can see that PAF, lysoPAF and lysolecithin mixed with ovolecithin in the fluid state (at room temperature) act as perturbants of the flatness of the bilayer plane (Fig. 4B), but that only little alteration of the local order parameter (Fig. 4A) is seen. This holds except at very high concentrations of lysoderivates, when fully interdigitated lamella are supposed to constitute a very narrow frame hindering the motion of an extended part of the probes. The useful distinction between the orientation distribution of the local directors of the lamellar phase and the motion of the acyl chains inside the lamella is depicted in Figure 1 and is only available after computer simulation of spectra from oriented smectic liquid crystals.

Figure 5 displays the ^{31}P NMR (202 MHz) spectral alterations for oriented samples (vertical concentric lamellae in a vertical magnetic field) containing 30% and 50% of PAF mixed with ovolecithin at 25°C.

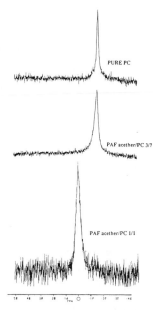

Figure 5: ^{31}P NMR spectra (202 MHz) of oriented multibilayers of ovolecithin enriched with PAF. The concentric lamellae are oriented parallel to the magnetic field. The perpendicular component of the Chemical Shielding tensor of ^{31}P is altered on the spectra when the motions of the phospholipids are out of the plane of the parallel lamellae (see Fig. 2).

The unique resonance line for 0 and 30% PAF at 15 ppm upfield from the resonance of an isotropic suspension corresponds to the perpendicular component (since the molecular axis is perpendicular to the magnetic field we detect the perpendical component) of the chemical shielding tensor whose motionally averaged anisotropy is -45ppm. This is to say that: i) the resonances of ^{31}P in PAF and in lecithin are undistinguishable at 202 MHz, ii) that no micellization has taken place and iii) that a local director of the phase is still observable. Figure 2 indicates the disturbance of the lamellae which has been simulated to take into account the increased linewidth of the line at 30% PAF (550 Hz instead of 300 Hz in the absence of perturbant) without a significant shift of the line position. At 50% PAF the line is not only considerably broadened (870 Hz), but it has also

shifted to the isotropic position indicating a phase
alteration with a apparent loss of the local director for the phospholipid
motion in the 202 MHz NMR time-scale. This is due to the lateral diffusion
on a non planar surface with a rate much faster than the chemical shielding
tensor (in frequency unit, 6.10^4 rad/sec at 202 MHz)

Three different models of reorientation of the phospholipid molecules have
been detailled (Olivier J.L. et al.) where the ratio of the lateral
diffusion constant ($D_L=10^{-11}$ $m^2.s^{-1}$ according to the lit erature) to the
average radius of curvature of the putative phase are challenged with the
correlation time for the phospholipid motion estimated from the linewidth
on the NMR spectra.

The 3 models are i) the mosaic structure (so-called the "ice floe" where
randomly oriented domains are delineated by abrupt edges) ii) the rippled
structure (so-called the "corrugated iron") iii) PAF/lecithin comicelles
adhering to the glass surface.

Only the rippled structure, with a continuous change of the molecular
orientation approximated by a Brownian motion of correlation time
$T_R=<R^2/6D_L>$ (R= root mean square curvature radius of the bilayer= 21 nm) is
consistent with the observed linewidth and a resonance line position shift
smaller than the linewidth (Olivier, J.L. et al.). The other models
proposed release narrower linewidths (comicelles) and a shifted resonance
peak position.

Finally, X ray diffraction shows that perturbations sensed by the magnetic
resonance techniques as a disorientation of the local director of the
lamellar phase, are accompanying an interdigitation of the fatty acyl
chains.

From Figure 6, one can see the emergence of 3 orders of a lamellar phase
reflection with a repeat distance around 55 A. This repeat distance
coexists on the diffractogramms with the "normal" lamellar distance around
71 A.

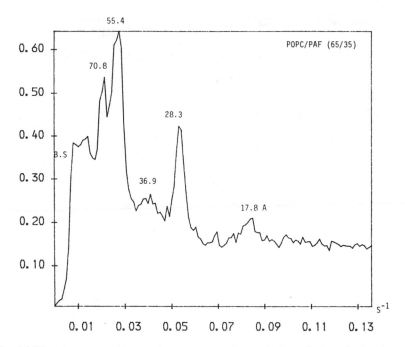

Figure 6: Diffractogram of a water suspension of 1 palmitoyl 2 oleoyl 3sn glycerophosphorylcholine enriched in PAF (35%) at 5°C. The X ray intensity is plotted as a function of the reciprocal spacing after a recording of the diffracted photons from the synchrotron beam during 10 times 2.5 sec (averaged diffractogram from 10 separated frames). The figure displays the small angle region from the beam center.

Figure 7 takes into account the observations released after ESR, NMR and X-ray studies conducted on lecithin films enriched in asymmetric phospholipids, bioactive (PAF) or non-bioactive (lysoderivates). Figure 7 is essentially a cartoon of the segregation of PAF into a blister delineated by an interdigitated lamella embedded within the "normal" lecithin bilayer, and which results in the disorientation of the planar alignment of the phopholipids polar heads, this is to say the local director of the lamellar phase. The stacking and the arrangement of such altered membranes could result in the emergence of the repeat distances quoted in Figure 7.

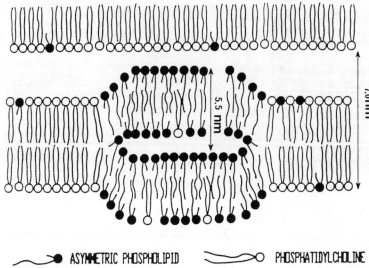

Figure 7: Model of the "blisterized" cluster of PAF in a lecithin host bilayer. The blister formed after segregation of the asymmetric lipid is thin-walled by an interdigitated lamellæ which perturbs the flatness of the host lamellae.

After the preceeding results, no clear evidence for a differential behaviour of bioactive and non bioactive lipids is given. On the contrary, the different asymmetric phospholipids share the same efficiency to disrupt the steady organization of the membrane bilayer. Then,at the time of the cell activation, the lipid mediator PAF would have its biophysical activity enhanced by lysoderivates which occurs after the stimulation of phospholipase A2. The activity of diglycerides which also appear briefly (before conversion to phosphatidic acid) during the cell activation, has to be challenged.

The activity of asymmetric phospholipids is restricted by the temperature,
diglycerides at low concentration and cholesterol.

Temperature prevents the segregation of the asymmetric species and
cholesterol forms a stable complex with lysoderivates (Rand, R.P. et al.).
Following this line, we have checked how the structural activity of PAF
released during cell activation could have been modulated.

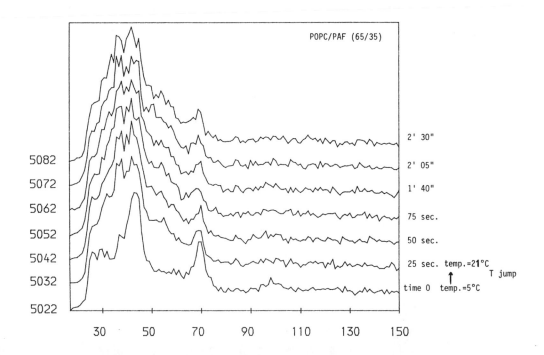

Figure 8: Sequential X ray diffractograms of 1 palmitoyl-2 oleoyl-
phosphatidylcholine/platelet activating factor (POPC/PAF: 65/35) liposomes
(20 mg) recorded after a T-Jump from 5 to 21°C. Each frame is recorded
after a 2.5 sec exposure to the synchrotron beam, the frame number is
indicated on the left margin and the corresponding time elapsed after the T
Jump on the right margin.

Figure 8 displays the diffractograms recorded sequentially after a T-jump from 5°C to 21°C. This final temperature is over the lamellar to micellar critical temperature (19°C) ascribed for pure PAF by Huang, C. et al. (1986). However it is below the transition of the host phospholipid (around 29°C) according to our DSC measurements (see below Figure 12). At this intermediary temperature, the interdigitated lamellae traced by the second order reflection of the phase at 28 A are detectable up to 2.5 minutes after the T-jump. Figure 9 shows that at this intermediary temperature, the interdigitated lamellae give still proeminent diffraction spots at 58.8 A and 28.4 A (2nd order), the contribution of the "normal" bilayer at 77 A being increased after comparison with the 5°C diffractogram.

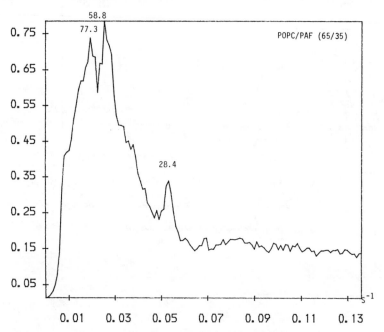

Figure 9: Diffractogram of POPC/PAF (65/35) liposomes at 21°C.

When the temperature is increased further (above the Tc of the host POPC) or when the asymmetric lipid is added to low-temperature melting host phospholipids (DOPC or ovolecithin), one can see a fusion of the reflections into a broad band centered on 59 A. Inside this broad reflection, the loss of resolution does not allow to confirm or to rule out the coexistence of "normal" and interdigitated bilayers.

The addition of diglycerides (5% DODG) to PAF (35%) or to lysolecithin (35%) in POPC (60%) does change the diffratogram. It reduces considerably the relative amount of interdigitated lamellae (Figure 10). The same observation can be made after addition of DODG (5%)to lysolecithin enriched liposomes or after the addition of DPDG (5%) (data not shown).

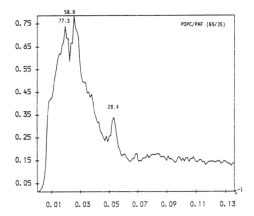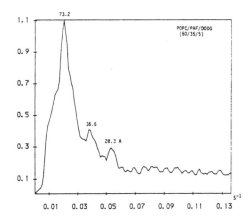

Figure 10: X ray diffractograms showing the influence of diglycerides at a low molar ratio (DODG 5%) on the interdigitated phase formed by asymmetric phospholipids (PAF 35%) mixed within an host lecithin (POPC 60%). The temperature is kept for both samples at 21°C, but the diffractogram of DODG/PAF/POPC liposomes is remarkably constant as a function of the temperature (between 5 and 21°C)(not shown).

Differential Calorimetry scans confirm this strong influence of a low concentration of diglycerides upon the emergence of the interdigitated lamellae promoted by asymmetric phospholipids. In Figure 11 one can see that DODG 5% erases most of the heat absorption related to PAF (around 20°C). It also acts to smooth out the residual heat absorption as a function of the number of heating scans performed on the sample, this is to say that it helps the dislocation of the asymmetric phospholipid clusters.

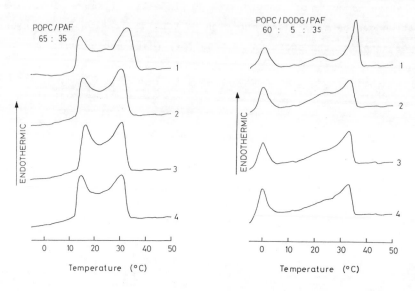

Figure 11: DSC of liposomes of POPC/PAF (65/35) and POPC/PAF/DODG (60/35/5). The thermograms were measured in a Perkin Elmer DSC-4 calorimeter and scanned at 4°C/min. The number indicates the scan order.

Cholesterol is recognized to counteract the structural activity of asymmetric phospholipids. PAF, like lysoderivatives, has its effects counteracted stoichiometrically by cholesterol in multibilayers probed by ESR (Figure 12). Observations by X-ray (not shown) give indications of a strongly ordered lamellar phase with a first order repeat distance indexing on 66 A. It is manifest at low or high temperature that the thermal stability of the complex formed between cholesterol and the asymmetric lipid is very high (see Olivier, J.L. et al.).

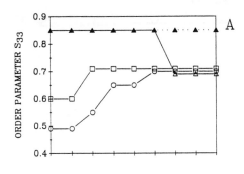
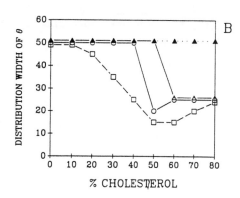

% CHOLESTEROL

Figure 12: Parameters calculated to fit the ESR spectra of 5 nitroxide
stearic acid probing oriented ternary mixtures of asymmetric phospholipid,
cholesterol, and egg lecithin. The values of the spectroscopic parameters
(order parameter and orientation distribution width of the local director)
were obtained from simulated spectra as described in Fig.1 and 4. PAF (○
), lysoPAF (△) and lysoPC (□).

According to the preceeding results the structural activity of PAF and
asymmetric phospholipids is restricted by the temperature, by the
diglycerides secreted during the cell activation and by the cholesterol
present in the biomembrane.

However, the possibility that lysoderivatives cooperate positively with
PAF still holds.

The presence of natural constitutive asymmetric phospholipids in biomembranes has been traced in rat platelets.

Preliminary studies suggest a role for these lipids, which would be substrates of the activated phospholipase C.

Figure 13 displays the molecular species of diglycerides (DG) secreted 30 seconds after the addition of 2U/ml thrombin to a rat platelets suspension. At this early stage, the secretion of DG has been shown to peak, before the conversion of DG into phosphatidic acid.

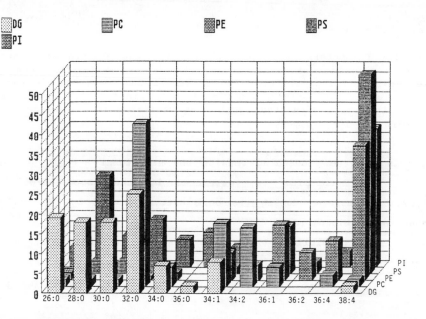

Figure 13: Molecular species composition of the diglycerides and of the various phospholipid classes in stimulated rat platelets. Each horizontal line represents a particular lipid and each column a particular molecular specie assayed by chemical ionization mass spectrometry of the benzoylated derivative (the polar head of the phospholipid being enzymatically hydrolysed and the glycerol hydroxyl group derivatized). The number in front of the column indicates the total fatty acyl chains carbon number and double bonds (for instance, 32:0 suggests 16:0-16:0). The quantification is performed after integration of each particular ion current corresponding to a particular mass (for instance, integration of the ion current corresponding to mass 690 releases the amount of 32:0).

The procedure of desorption chemical ionisation mass spectrometry (DCIMS) has been previously challenged successfully with the HPLC procedure by Masrar, H. et al.. Chemically pure phospholipids release mass spectra with no indication of fragmentation of the fatty acyl chains. One must be awared that the reliability of the procedure is satisfactory if the mass spectra recording (including the reagent gas pressure in the ionization source) as well as the thermal desorption rate (+ 5 mA/sec) are kept constant.

In the pattern presented, the molecular species of diglycerides differ from any possible single parent phospholipid class. DG are enriched in short acyl chains (26, 28, 30 C for both acyl chains, which means for instance 16:10, 16:12 or 18:10, 16:14 or 18:12). These short chains diglycerides hypothetically could derivate from precursor phospholipids with a high chains inequivalence. The secreted diglycerides are deprived in arachidonic acid (20:4) as regard to PE, PS and PI which contain over 40% of 36 C (16-20:4)or 38 C (18-20:4) with 4 double bonds. The stri king high content in 32 C (16:0-16:0) for diglycerides or PC has been temptingly explained for rat platelets by CoA dependent transacylation reactions of lysophosphatidylcholine (Masrar, H. et al.).

A possible precursor for the asymmetric molecular species would be PI which contains a significant level of short chain species and is a recognized substrate for phospholipase C generating the diglycerides. Under that circonstance, one may hypothesize that such short chain PI with a probable significant acyl chains inequivalence, is a favourable substrate of phospholipase C. It will be also a favourable substrate of phospholipase A2, which would leave little arachidonic acid in the diglycerides secreted after the hydrolysis by the competitive pathway of phospholipase C.

The precise role in the enzyme affinity for the polar heads out of alignment as suggested from the data obtained after a physical studies of highly asymmetric lipids, such as PAF or lysoderivates, remains opened to question.

Figure 13 represents the data from a particular experiment but other experiments, though not similar, support also the early occurence of diglycerides enriched in short chains. The very low amount of diglycerides, which peak early and briefly after the cell activation, has prompted us to study currently the molecular species of the phosphatidic acid, a accumulating lipid class after thrombin stimulation (to be published).

DISCUSSION

PAF displays a structural activity toward the phospholipid bilayer which is identical with non-bioactive lysoderivatives (lysoPAF or lysolecithin). No clear cut qualitative or quantitative distinction is possible after studies with ESR, ^{31}P NMR or X ray diffraction, which suggests a little influence of the hydroxyl instead of the acetyl group on the behaviour of the asymmetric lipids in membrane.

The activity of the asymmetric lipids is related to its ability to self aggregate and to constitute domains in the membrane with an extended interdigitation of the acyl chains. This, in turn, appears to disalign the phospholipid head groups and to disorient the local director of the plane bilayer.

When the temperature is above the gel-liquid transition temperature of the host lipid, the asymmetric lipid segregation is partially impeded and its structural activity is apparently decreased but one cannot exclude the occurence of short-lived small clusters of asymmetric lipids in agreement with the findings of spectroscopical studies. The asymmetric phospholipid would be distributed according to its affinity to self aggregate and to its propensity to lateral diffusion into the surrounding fluid lipid.

Diglycerides, at low (5%) concentration, induce the same effect as a temperature rise. Cholesterol counteracts PAF self aggregation by forming a stable association with the asymmetric lipid to give rise to a thermostable lamellar arrangment with a long repeat distance.

From a molecular analysis of diglycerides secreted after rat platelets stimulation by thrombin, a biological relevance for asymmetric molecular phospholipid species is proposed. These asymmetric species would be a favorite substrate to phospholipase C which generates the diglycerides and to phospholipase A2 which release arachidonic acid. This is hypothesized from the molecular species analysis by mass spectroscopy of diglycerides enriched in short acyl chains and deprived in arachidonic acyl moiety.

Acknowledgements: We are grateful to Dr. J.C. Gomez Fernandez at the Universidad de Murcia (Spain) for his help providing the DSC data. The expertise of F. Chevy at CHU St Antoine with DCIMS is gratefully acknowledged.

REFERENCES

Blank, M.L., Robinson, M., Fitzgerald, V. and Snyder, F. (1984) J. Chromatogr. 289, 473-482.

Braquet, P. (1987) Drugs of the Future, 12(7), 643-699.

Chachaty, C. (1985) J. Chim. Phys. 82, 621.

Chachaty, C., Quaegebeur, J.P., Caniparoli, J.P., and Korb, J.P. (1986) J. Phys. Chem. 90, 1115-1122.

Chachaty, C., Soulie, E. and Wolf, C. (1991) J. Chim. Phys. 88, 153-172.

Ginka, K.G., St. Denney, I.H. and Nemecek, G.M. (1986) in: New Horizons in PAF Research (C.M. Winslow, M.L. Lee Eds) New York.

Honda, Z., Nakamura, M., Miki, I.,Minami, M., Watanabe, T., Seyama, Y., Okado, H., Toh, H., Ito, K., Miyamoto, T. and Shimizu, T. (1991) Nature 349, 342-345.

Huang, C and Levin, I.W. (1983) J. Phys. Chem. (1987) 1509-1530.

Huang, C., Mason, J.T., Stephenson, F.A. and Levin, I.W. (1984) J. Phys. Chem. 88, 6454-6458.

Huang, C., Mason, J.T., Stephenson, F.A. and Levin, I.W. (1986) Biophys. J. 49, 587-595.

Hwang, S.B., Lee C.S.C., Cheah, M.J. and Shen, T.Y. (1983) Biochemistry 22(4), 4756-63.

Killian, J.A., Borle, F., De Kruijff, B. and Seelig, J. (1986) Biochim. Biophys. Acta 854, 133-142.

Kramp, W., Pieroni, G., Pinkard, R.N., and Hanahan, D.J. (1984) Chem. Phys. Lipids 35, 49-62.

O'Flaherty, J.T. (1987) in: PAF and Related Lipid Mediators (F. Snyder Ed) Plenum Press, New York, 283.

Olivier, J.L., Chachaty, C., Quinn, P.J. and Wolf, C. (1991) J. Lipid Mediators 3, 311-332.

Olivier, J.L., Chachaty, C., Wolf, C. and Bereziat, G. (1988) Biochimie 71(1), 105-111.

Olivier, J.L., Chachaty, C., Wolf, C. and Bereziat, G. (1989) Agents and Actions 26(1-2), 121-122.

Masrar, H., Bereziat, G. and Colard, O. (1990) Arch. Biochem. Biophys. 281, 116-123.

Peplow, P.V. and Mikhailidis, D.P. (1990) in: Prostaglandins Leucotrienes and Essential Fatty Acids 41, Longman Group UK, pp. 71-82.

Presscott, S.M., Zimmerman, G.A. and Mc Intyre, T.M. (1990) J. Biol. Chem. 265, 17381-84.

Rand, R.P., Pangborn, W.A., Purdon, A.D. and Tinker D.O. (1975) Can. J. Biochem. 53, 189-195.

Sawyer, D.B. and Andersen, O.S. (1989) Biochim. Biophys. Acta 987, 127-132.

Schneider, D.J. and Freed, J.H. (1989) in: Biological Magnetic Resonance, L.J. Berliner Edr., vol. 8, Plenum Press, New York, pp. 1-76.

Thuren, T., Virtanen, J.A. and Kinnunen, P.K.J. (1990) Chem. Phys. Lipids 53, 129-139.

Vallari, D.S., Austinhirst, R. and Snyder, F. (1990) J. Biol. Chem. 265, 4261-4265.

Van Echteld, C.J.A., De Kruijff, B., Mandersloot, J.G. and De Gier, J. (1981) Biochim. Biophys. Acta, 649, 211-220.

Wu, W.G., Huang, C., Conley, T.G., Martin, R.B. and Levin, I.W. (1982) Biochemistry 21, 5957-5961.

Progress in Membrane Biotechnology
Gomez-Fernandez/Chapman/Packer (eds.)
© 1991 Birkhäuser Verlag Basel/Switzerland

MEMBRANE STABILIZATION BY ANTIOXIDANT RECYCLING

by Dr. Lester Packer

Department of Molecular and Cell Biology, 251 Life Sciences Addition, University of California, Berkeley, CA 94720, USA

SUMMARY: Enzymatic and non-enzymatic pathways of lipophilic antioxidant action can be exploited to stabilize membranes. Vitamin E can be enzymically regenerated from its chromanoxyl radicals formed in the course of peroxidation. Chromanoxyl radicals of vitamin E can be reduced by NADPH-, NADH-, and succinate-dependent electron transporting enzymes of microsomes and mitochondria. Ubiquinones can act as co-factors in the enzymic regeneration of vitamin E. Other aqueous reductants (ascorbate, glutathione, dihydrolipoate) can synergistically enhance vitamin E regeneration. Thus, vitamin E molecules (tocopherols and tocotrienols) possess a unique ability to act as membrane free radical harvesting centers due to their regeneration in membranes.

Non-enzymatic pathways of antioxidant action can be exemplified by the thioctic acid/dihydrolipoic acid redox couple (TA/DHLA) in liposomes and LDL. TA has been found to confer protection in tissues and membranes against oxidative damage. TA after it is absorbed, is reduced to become active as an antioxidant. DHLA acts a a "double-edged sword" in that it appears to interact directly with peroxyl radicals in the membrane or indirectly to reduce tocopheroxyl radicals via a cascade mechanism involving the reduction of ascorbate, which in turn reduces tocopheroxyl radicals to tocopherol. This mechanism of the TA/DHLA couple works in recycling of vitamin E both in membranes and in low density lipoproteins (LDL) where it acts to stabilize them against free radical mediated oxidation of lipids and protein.

INTRODUCTION

Living cells have three main systems for protection and repair under oxidative stress: (i) direct antioxidant enzymes (SOD, catalase, peroxidases), (ii)

proteases and phospholipases activated by oxidative modification of membranes, and (iii) lipid- and water-soluble antioxidants. (i) and (ii) constitute the two lines of enzymic antioxidant defense, while (iii) has been considered in the past to be mainly non-enzymic and dependent on nutritional factors. Our recent studies show that this is not the case: both non-enzymatic and enzymatic exist and knowledge of these mechanisms can be exploited to stabilize membranes.

Vitamin E is believed to be the major lipid-soluble chain-breaking antioxidant in membranes (Packer & Landvik, 1989). However its membrane concentration is usually lower than 0.5 - 0.1 nmoles/mg of protein or 0.05 - 0.1 mole% of membrane phospholipids. The rates of lipid radical generation dependent on electron transport (e.g. NADPH-dependent lipid peroxidation in liver microsomes) may be as high as 1 - 5 nmoles/mg of protein per minute. Nevertheless, under physiological conditions the low concentration of vitamin E is sufficient to prevent membrane oxidative damage. We have demonstrated that vitamin E can be enzymically regenerated from its chromanoxyl radicals formed in the course of peroxidation (Hiramatsu & Packer, 1991; Hiramatsu, et al., 1990; Kagan, Packer, et al., 1990; Kagan, Serbinova, & Packer, 1990b; Kagan, Serbinova, & Packer, 1990c; Maguire, et al., 1989; and Packer, Maguire, et al. 1989). ESR and HPLC data show that chromanoxyl radicals of vitamin E can be reduced by NADPH-, NADH-, and succinate-dependent electron transporting enzymes of microsomes and mitochondria (ubiquinones act as co-factors in enzymic regeneration of vitamin E). These enzymes involved in the regeneration of lipid-soluble phenolic antioxidants constitute the third line of antioxidant defense, controlled by the cell genome.

In this article, human low density lipoproteins, which can be considered to be a natural counterpart for liposomes, are used to illuminate the action of antioxidant recycling mechanisms in the protection and stabilization of lipid-protein systems.

MATERIALS AND METHODS

ISOLATION OF LDL: (Kagan, Freisleben, et al., 1991; and Kagan, Serbinova, Forte, et al., 1991): A pool of fresh plasma from normolipidemic subjects was used for isolation of LDL. The plasma was adjusted to d 1.019 g/ml with solid NaBr and centrifuged 40,000 rpm, in a 50.3 Beckman rotor for 24 hr at 4ºC.

The top 1 ml was harvested by aspiration and discarded. The infranatant was then adjusted to d 1.063 g/ml and centrifuged an additional 24 hr after which the top 1 ml representing LDL (d 1.019-1.063 g/ml) was collected by aspiration. The LDL were dialyzed extensively against phosphate buffer (50 mM, pH 7.4) prior to use.

FLUORESCENCE SPECTRA: Fluorescence emission spectra of LDL suspensions in phosphate buffer (50 mM, pH 7.4, 0.2 mg protein/ml) were recorded in the spectral region 400-500 nm and excitation at 365 nm (slit widths were 6 nm), a maximum that is specific for oxidatively modified LDL, in accordance with the procedure described by Steinbrecher.

GENERATION OF CHROMANOXYL RADICALS: Chromanoxyl radicals from α-tocotrienol, α-tocopherol and its homologue were generated (Serbinova, E., Kagan, V., Han, D., and Packer, L., 1991) using: 1) an enzymic oxidation system (soybean 15-lipoxygenase + linolenic acid), 2) hydrophobic azoinitiator of peroxyl radicals, 2,2'-azo-bis(2,4-dimethylvaleronitrile) (AMVN), and 3) UV-irradiation. When the enzymic oxidation system was used the incubation medium (100 μl) contained LDL (10 mg protein/ml) in 50 mM phosphate buffer, pH 7.4 at 25°C. The concentration of exogenously added chromanols was 80 nmoles/mg protein. Linolenic acid (1.4 mM) +lipoxygenase (10 U/μl) and chromanols were subsequently added to LDL suspension. Chromanols were added in ethanolic solution. With the azo-initiator the incubation medium was essentially the same but AMVN (5.0 mM) was added instead of (lipoxygenase + linolenic acid) and the reaction was carried out at 40°C.

IRRADIATION: Irradiation was by a solar simulator (Solar Light Co., model 14S), whose output closely matches the solar spectrum in the wavelengths 290-400 nm. The samples were illuminated directly in the ESR resonator cavity; the distance between the light source and the sample was 30 cm. The power density of the light at the sample surface in the spectral region 310-400 nm was 1.5 mW/cm^2 and dropped to 10% of this value at 290 nm.

ESR MEASUREMENTS: ESR measurements were made on a Varian E 109E spectrometer in gas permeable Teflon tubings (0.8 mm internal diameter, 0.013

mm thickness obtained from Zeus Industrial Products, Raritan N.J. USA). The gas permeable tube (approximately 8 cm in length) was filled with 60 μl of a mixed sample, folded into quarters and placed in an open 3.0 mm internal diameter EPR quartz tube such that all of the sample was within the effective microwave irradiation area. ESR spectra were recorded either in the dark or under continuous UVAB irradiation by the solar simulator in the ESR cavity. Spectra were recorded at 100 mW power and 2.5 gauss modulation, and 25 gauss/minute scan time. Spectra were recorded at room temperature under aerobic conditions by flowing oxygen through the ESR cavity. Chromanoxyl and ascorbyl radical ESR signals were recorded at 3245 gauss magnetic field strength, scan range 100 gauss, and time constant 0.064 sec.

HPLC MEASUREMENTS OF α-TOCOPHEROL CONTENT: α-Tocopherol was assayed by reverse phase HPLC using a C-18 column (Waters, Inc.) with an in-line electrochemical detector. The effluent was methanol:ethanol 1:9 (v/v), 20 mM lithium perchlorate. Tocopherol was extracted into hexane from sodium dodecyl sulfate-treated samples as described earlier (Serbinova, et al., 1991).

RESULTS AND DISCUSSION

THE VITAMIN E CYCLE AND ITS INTERACTION WITH WATER- AND LIPID-SOLUBLE ANTIOXIDANTS BY ENZYMATIC AND NON-ENZYMATIC PATHWAYS: We have studied the vitamin E cycle and characterized it in four different systems. Lipid liposomes like dioleolyl phosphytalcholine liposomes [DOPC], natural membranes like rat liver inner mitochondrial (Kagan, Serbinova, & Packer, 1990c; Maguire, et al., 1989; and Packer, Maguire, et al., 1989) and microsomal membranes (Mehlhorn, et al., 1989 and Serbinova, et al., 1991), and human low density lipoproteins (Kagan, Freisleben, et al., 1991; and Kagan, Serbinova, Forte, et al., 1991). The vitamin E cycle is shown in general terms in Figure 1. Depicted on the left are various mechanisms whereby membranes can be oxidatively stressed, to generate chromanoxyl radicals from vitamin E. UVB irradiation directly interacts with vitamin E molecules, which absorb light in this region to generate chromanoxyl radicals. Superoxide ions generated by various electron transport systems in the cell are other sources for oxidizing vitamin E to its radical form. In addition,

144

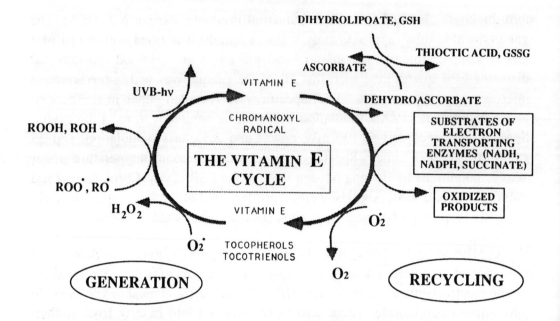

Figure 1: The Vitamin E Cycle--synergistic action with water-soluble and lipid-soluble antioxidants by enzymatic and non-enzymatic mechanisms.

various artificial and natural mechanisms exist for generating lipid radicals (alkoxyl, peroxyl radicals) which react with vitamin E to generate the vitamin E radical, a more persistent radical which can be readily detected by electron spin resonance techniques. The steady-state level of tocopheroxyl or tocotrienoxyl radicals are then influenced by the absence or presence of other reductants (Kagan, Serbinova, Forte, et al., 1991; Kagan, Serbinova, & Packer, 1990c; and Serbinova, et al., 1991), both within the membrane and externally, which can interact at the membrane interface to quench the radicals.

Non-enzymatic pathways of regeneration of vitamin E are afforded by water-soluble reductants such as vitamin C, ascorbic acid. Figure 2 shows an example where suspensions of human LDL were exposed to an enzymatic oxidation system comprised of lipoxygenase and linolenic acid, which generates peroxyl radicals of linolenate (Kagan, Freisleben, et al., 1991 and Kagan, Serbinova, Forte, et al., 1991). These, in turn, react with vitamin E. In this case α-c-6, a homologue of vitamin E which contains a shorter side chain than the natural tocopherol, was used (Kagan, Serbinova, & Packer, 1990c). One can observe the pentameric ESR signal, characteristic of vitamin E. Then, ascorbic acid is added and, as can be seen, the vitamin E radical signal disappears and the new ESR signal is dominated by the semi-ascorbyl radical , which is a large prominent signal seen in the center of the trace. After about 30 minutes of incubation time, the ascorbate which has been added, continuously regenerating the vitamin E, eventually is expended, and when it disappears, the vitamin E free radical signal reappears. It can be observed that the signal that reappears is somewhat smaller in intensity than the original signal, suggesting that radical-radical interactions between vitamin E chromanoxyl radicals have occurred to cause its consumption [loss during the incubation period].

In natural membranes, in addition to ascorbate, there are enzymatic pathways of quenching chromanoxyl radicals. Substrates of electron transport in mitochondrial and microsomal membranes, such as succinate, NADPH, and NADH are examples. These appear to be mediated through the action of ubiquinones. We have demonstrated that ubiquinone, while exerting an antioxidant action in its own right to quench lipid peroxidation (Kagan, Serbinova, Koynova, et al., 1990), in membranes where electron transport through ubiquinones exists, ubiquinones appear to be "a slave to vitamin E," by donating electrons to tocopheroxyl radicals (Kagan, Serbinova, & Packer, 1990a).

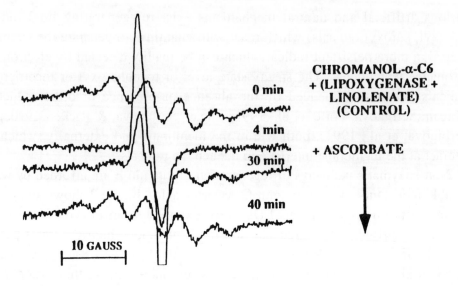

Figure 2: Interaction of tocopheroxyl and ascorbyl radicals in suspensions of human LDL. LDL were incubated with the a-tocopherol short chain homologue a-C-6. Exposure to the enzymatic oxidation system (lipoxygenase plus linolenic acid) causes formation of the tocopheroxyl radical signal (5 line spectrum). The ascorbyl radical signal develops after ascorbate is added (large central signal).

Ascorbate, glutathione, dihydrolipoate, and other reductants can synergistically enhance vitamin E regeneration. The well known effects of these physiologically important antioxidants (reductants) with vitamin E are mediated via their ability to transfer electrons necessary for recycling chromanoxyl radicals in membranes. Thus, vitamin E molecules (tocopherols and tocotrienols) possess a unique ability to act as membrane free radical harvesting centers due to their enzymic and non-enzymic regeneration in membranes. In this regard, α-tocotrienol (having an unsaturated isoprenoid side chain) as compared to α-tocopherol has been found to possess 40 - 60X greater antioxidant potency. Three factors seem to explain this remarkable effect: greater recycling activity, more homogenous membrane distribution, and greater membrane mobility (Serbinova, et al., 1991).

The remarkably increased antioxidant potency of α-tocotrienol has been observed in all four of the systems that we have been studying. Figure 3 gives an example of a comparison the the antioxidant recycling activity of d-α-tocopherol as compared to d-α-tocotrienol and the α-C-6 chromanol homologue. The most effective form of vitamin E is the short chain homologue. This is followed by α-tocotrienol and α-tocopherol, which is the poorest. Although these results were made with oxidatively stressed LDL, similar results have been found in microsomal and mitochondrial membranes and DOPC liposomes.

The thioctic acid/dihydrolipoic acid couple (TA/DHLA) is a unique system. Normally lipoamide exists as the co-factor of α-keto-dehydrogenases and is covalently bound in animals. Thus, its presence is at the level of micronutrient. However, TA, which can readily be fed to animals, has been found to confer protection in tissues and membranes against oxidative damage. This is believed to occur because TA after it is absorbed, is reduced enzymatically or non-enzymatically and then becomes active as an antioxidant. DALA acts a a "double-edged sword" in that it appears to interact directly with the membrane to reduce tocopheroxyl radicals (weak effect) or to reduce ascorbate which in turn acts at membranes to reduce tocopheroxyl radicals (stronger effect). This mechanism of the TA/DHLA couple works in recycling of vitamin E both in membranes and in low density lipoproteins (LDL) where it acts to stabilize them.

SYNERGISTIC ACTION OF DHLA WITH ASCORBATE IN LDL VITAMIN E RECYCLING ACTIVITY:
The synergistic action of dihydrolipoic acid in

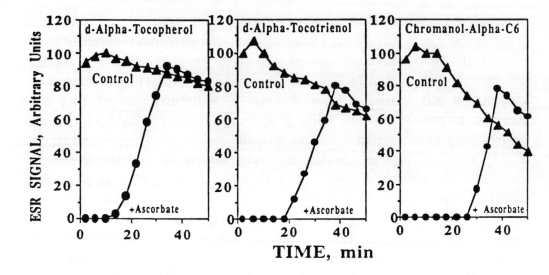

Figure 3: Time course of the lifetime of chromanoxyl radical signals in the absence and presence of ascorbate. Comparison of recycling activity by d-a-tocopherol, d-a-tocotrienol, and the d-a-C-6 short chain vitamin E homologue.

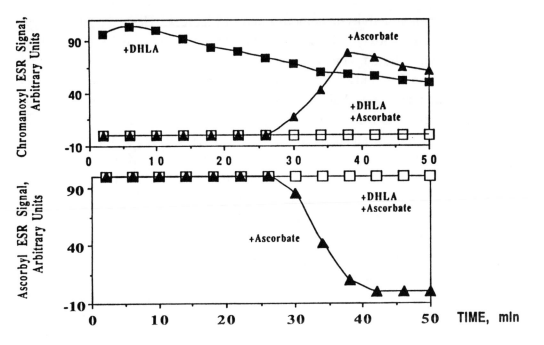

Figure 4: Reactions of chromanoxyl and ascorbyl radicals in human LDL exposed to oxidants formed by an enzymatic oxidation system comprised of lipoxygenase and linolenic acid.

vitamin E recycling activity is demonstrated in Figure 4. Here, the steady state levels of the chromanoxyl radical signals of vitamin E and of the semi-ascorbyl radical are demonstrated. In the presence of DHLA, the chromanoxyl radical signal of vitamin E is only slightly reduced over time, whereas if ascorbate is added, the vitamin E radical signal is quenched until the ascorbate is exhausted. So, in this experiment after 25 minutes, when ascorbate is expended, the vitamin E chromanoxyl radical signal returns. If, however, DHLA and ascorbate are combined together, as can be seen, no reappearance of the vitamin E chromanoxyl radical signal is observed. The lower part of the experiment demonstrates that this is indeed an effect on maintenance of the semi-ascorbyl free radical signal. In the presence of DHLA and ascorbate, the vitamin C radicals signal persists, whereas in the absence of DHLA, after about 25 minutes, the ascorbyl signal begins to disappear. No disappearance of the ascorbyl signal is seen even up to 50 minutes.

In order to prove that indeed these interpretations are correct, one can test for the loss of chromanols by using HPLC techniques. The chromanols will disappear mainly during the time period when they are in their free radical form, when they can undergo radical reactions with themselves or with other radicals to be lost from the system as active antioxidants. Thus it can be seen in Table I, that using two different oxidation systems, UVB light or lipoxygenase plus linolenic acid, it is possible to cause the destruction of endogenous vitamin E. The enzymatic oxidation system is particularly effective in this regard.

Hence, during the time course of the incubation system with LDL, more than half of the vitamin E is lost if the LDL is incubated in the presence of ascorbate, and 90% is lost if it is incubated in the presence of DHLA. But if both ascorbate and DHLA are present, only 30% of the vitamin E is lost i.e. a synergistic effect. These results are also reflected by fluorescent studies, which are indicative of the accumulation of end products of lipid peroxidation. As can be seen, large increases in the accumulation of fluorescent products is observed under conditions where lipoxygenase plus linolenic acid are incubated without either ascorbate or DHLA, whereas ascorbate or DHLA both largely protect against the accumulation of fluorescence products.

Based on these results, a scheme for how the thioctic acid/dihydrolipoic acid couple works was constructed and is illustrated in Figure 5. This scheme demonstrates that after animals are fed thioctic acid, the reduction of the thioctic

151

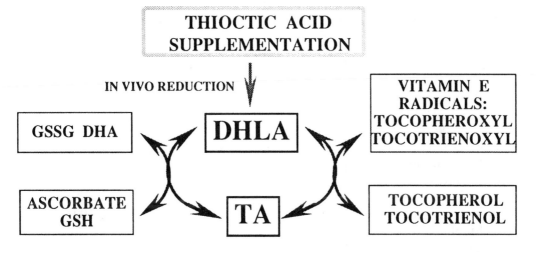

Figure 5: Thioctic acid/dihydrolipoic acid couple--a double-edged sword. mediator between water- and lipid-soluble antioxidant defenses.

152

Table I: Changes in the chromanol content and oxidation level (fluorescence intensity) of LDL by different oxidation systems and in the presence of different reductants.

CONDITIONS	CHROMANOLS % OF CONTROL	FLUORESCENCE INTENSITY % OF CONTROL
ENDOGENOUS VITAMIN E*		
CONTROL (50 min incubation)	100	100
+ UVB	84.0 ± 6.0	98.0 ± 6.0
+Lipoxygenase +Linolenic Acid	<0.3	330.0 ± 32.0
+Lipoxygenase +Linolenic Acid +Ascorbate	40.0 ± 3.0	97.0 ± 6.5
+Lipoxygenase +Linolenic Acid +DHLA	10.0 ± 1.0	120.0 ± 10.0
+Lipoxygenase +Linolenic Acid +Ascorbate+DHLA	70.0 ± 5.0	98.0 ± 8.0

*The concentration of endogenous vitamin E in LDL was 3 nmoles/mg protein. Incubation medium (100µL) contained: LDL (12 mg protein/ml) in phosphate buffer, pH 7.4 at 25°C. Additions: lipoxygenase (3U/µL) + linolenic acid (1.4 mM) ascorbate (1.5 mM), ascorbate (1.5 mM), DHLA (2 mM).

acid occurs *in vivo,* either by enzymatic or non-enzymatic pathways, or both to form the antioxidant active form dihydrolipoic acid, DHLA. This then, effectively is able to reduce water-soluble thiols such as glutathione and dehydroascorbic acid, to form reduced glutathione and ascorbate, respectively. DHLA can also react directly, although it is a more weak interaction, with vitamin E radicals to regenerate vitamin E to its natural form [tocopherol or tocotrienol]. DHLA acts most effectively when it works synergistically in conjunction with water-soluble reductants, such as ascorbate.

Membrane technology is benefitting from the use of disulfide and thiol absorption, such as to gold surfaces (Fabianowski, et al., 1989). Such molecules have the ability to self-assemble and to form coherent thin film barriers. David W. Grainger [this volume] is pioneering this kind of research. Such thin films have usefulness in electron transfer on metal surfaces (Li & Weaver, 1984). Though this is a much more specialized application of the use of thiols, their action and stabilizing membranes through biochemical and chemical mechanisms, by virtue of their antioxidant properties, are of equal interest in stabilizing membranes susceptible to lipid and protein oxidations.

Acknowledgements: This research was supported by NIH CA 47597, and ASTA Pharma [Frankfurt, Germany].

REFERENCES

Fabianowski, W., Coyle, L.C., Weber, B.A., Granata, R.D., Castner, D.G., Sadownik, A., and Regen, S.L.: Spontaneous assembly of phosphatidylcholine monolayers via chemisorption onto gold. Langmuir 5: 35-41, 1989.

Hiramatsu, M. and Packer, L.: Interactions of water- and lipid-soluble antioxidants with alpha-tocopheroxyl radicals in membranes. In: New Trends in Biological Chemistry, (Ozawa, T., ed.), pp. 323-331, Japan Sci. Soc. Press, Tokyo/Springer-Verlag, Berlin, 1991.

Hiramatsu, M., Velasco, R.D., and Packer, L.: Vitamin E radical reaction with antioxidants in rat liver. Free Rad. Biol. Med. 9: 459-464, 1990.

Kagan, V.E., Freisleben, H.J., Tsuchiya, M., Forte, T., and Packer, L.: Generation of Probucol radicals and their reduction by ascorbate and dihydrolipoic acid in human low density lipoproteins. Free. Rad. Res. Comm., 1991.

Kagan, V.E., Packer, L., Serbinova, E., Bakalova, R., Stoyanovsky, D., Zhelev, Z. , Harfouf, M., Kitanova, S., Rangelova, D.: Mechanisms of vitamin E control: lipid peroxidation and synergism and migration chelation interaction with hydroperoxides. In: International Symposium on Biological Oxidation Systems, (Reddy, Channa, ed.), Academic Press, Bangalore, India, October 1989, pp. 63, 1990.

Kagan, V.E., Serbinova, E.A., Forte, T., Scita, G., and Packer, L.: Recycling of vitamin E in human low density lipoproteins: interplay between water- and lipid-soluble antioxidants. Submitted.

Kagan, V., Serbinova, E., Koynova, G., Kitanova, S., Tyurin, V., Stoytchev, T., Quinn, P., and Packer, L.: Antioxidant action of ubiquinol homologues with different isoprenoid chain length. In: Biomembranes. Free Rad. Bio. Med., 9(2):117-126, 1990.

Kagan, V., Serbinova, E., and Packer, L.: Antioxidant effects of ubiquinones in microsomes and mitochondria are mediated by tocopherol recycling. Biochem. Biophys. Res. Comm., 169(3):851-857, 1990.

Kagan, V. E., Serbinova, E. A., Packer, L.: Generation and recycling of radicals from phenolic antioxidants. Arch. Biochem. Biophys., 280(1):33-39, 1990.

Kagan, V., Serbinova, E., Packer, L.: Recycling and antioxidant activity of tocopherol homologues of differing hydrocarbon chain length in liver microsomes. Arch. Biochem. Biophys., 282(2):221-225, 1990.

Li, T.T.T., and Weaver, M.J.: Intramolecular electron transfer at metal surfaces. Dependence of tunnelling probability upon donor-acceptor separation distance. J. Am. Chem. Soc. 106: 6107-6108, 1984.

Maguire, J.J., Wilson, D.S., and Packer, L.: Mitochondrial electron transport-linked tocopheroxyl radical reduction. J. Biol. Chem., 264:(36):21462-21465, 1989.

Mehlhorn, R.J., Sumida, S., and Packer, L.: Tocopheroxyl radical persistence and tocopherol consumption in liposomes and in vitamin E-enriched rat liver mitochondria and microsomes. J. Biol. Chem., 264(23):13448-13452, 1989.

Packer, L. and Landvik, S.: Vitamin E - introduction to its biochemistry and health benefits. In:Vitamin E Biochemistry and Health Implications, (A.T. Diplock, L.J. Machlin, Packer, L., W.A. Pryor, eds.) New York Academy of Sciences, Volume 570, pp.1-6, 1989.

Packer, L., Maguire, J.J., Mehlhorn, R.J., Serbinova, E., and Kagan, V.E.: Mitochondria and microsomal membranes have a free radical reductase activity that prevents chromanoxyl radical accumulation. Biochem. Biophys. Res. Comm., 159(1):229-235, 1989.

Serbinova, E., Kagan, V., Han, D., and Packer, L.: Free radical recycling and intramembrane mobility in the antioxidant properties of alpha-tocopherol and alpha-tocotrienol. Free Rad. Biol. Med., Vol. 10, pp. 263-275, 1991.

Progress in Membrane Biotechnology
Gomez-Fernandez/Chapman/Packer (eds.)
© 1991 Birkhäuser Verlag Basel/Switzerland

PURIFICATION AND SOME PROPERTIES OF E. coli α-HAEMOLYSIN

Helena Ostolaza, Borja Bartolomé*, Iñaki Ortiz de Zárate, Fernando de la Cruz* and Félix M. Goñi.

Dpt. of Biochemistry, University of the Basque Country, P.O. Box 644, 48080 Bilbao, Spain, and

*Dpt. of Molecular Biology, University of Santander Medical School, Santander, Spain.

SUMMARY: E. coli α-haemolysin has been purified following a simple procedure. The protein (110 kDa) exists in the form of large aggregates, held together mainly by hydrophobic forces. In the presence of urea or other chaotropic agents, the size of the aggregates decreases, while the specific activity is increased. α-Haemolysin releases the aqueous contents of phospholipid vesicles, but it does not promote intermixing of vesicle lipids.

α-Haemolysin is an extracellular E. coli protein that appears to be responsible for extraintestinal diseases such as infections of the human genito-urinary tract, meningitis, peritonitis, etc. (1,2). This toxin is also interesting from the point of view of biotechnology, since it is one of the few proteins extracellularly excreted by E. coli, therefore it can be used in the cell export of other gene products in the form of chimeric proteins (3).

α-Haemolysin is produced in an active form during the

exponential growth phase of the bacterium. At least four genes (hly A,B,C and D) are required for the production of the β-haemolytic phenotype (4). The product of gene hly C is involved, in an unknown way, in the intracellular activation of the main α-haemolysin polypeptide, i.e. the product of gene hly A (5), while genes hly B and D are related to protein transport.

Other interesting aspects of haemolysin, e.g. the mechanism of its haemolytic action, are virtually unknown at the molecular level. This is partly due to the fact that α-haemolysin purification has found a number of difficulties in the past: low haemolysin concentrations in the growth media, protein instability, etc. González-Carreró et al. (6) described the construction of an overproducing **E. coli** strain, containing the pSU 157 plasmid; they also proposed a purification procedure, and found that 6 M urea greatly increases the stability of α-haemolysin preparations. At present, a number of procedures for α-haemolysin purification are available, giving rise to similar, but not identical products, often poorly characterized from the physico-chemical point of view. This paper summarizes a series of experiments aimed at (a) finding an improved purification procedure, starting from the data by González-Carreró et al. (6), and (b) describing some properties of α-haemolysin, that may help to rationalize the present scattered, and sometimes contradictory, data. A preliminary report has been published elsewhere (7).

MATERIALS AND METHODS

Bacterial strains and plasmids: Strain RRI△M15 of E. coli K12 was the host for the plasmid used in this work. The re- combinant plasmid pSU 124, containing the hly genes, was obtained by cloning the hly determinant, from the Eco RI site at coordinate 9.5 kb to the Bgl II site at 20.1 kb in the map by Zabala et al. (8), in the Eco RI - Bam H1 sites of vector PUC 8.

Assay for haemolytic activity: Equal volumes of a 3% (v/v) suspension of washed sheep erythrocytes were added to serial twofold dilutions of α-haemolysin extracts in buffer (150 mM NaCl, 10 mM CaCl$_2$, 20 mM Tris-HCl, pH 7) (9). The mixtures were incubated at 37ºC for 45 min, then left at room tem- perature for a few hours, so that erythrocyte sedimentation occurred. The absorbance of the supernatants, appropriately diluted with distilled water, was read at 412 nm. The blank (zero haemolysis) consisted of a mixture of equal volumes of buffer and erythrocytes. Haemolytic activity (in haemolytic units (HU) x ml^{-1}) is defined as the dilution of α- haemolysin preparation producing 50% lysis of the erythro- cyte suspension.

Purification of α-haemolysin: E. coli (pSU 124) was grown on LB medium (10) in a shaking incubator, at 37ºC, to an A$_{550}$ ≈ 1.0. Cells were collected by centrifugation (13000 xg, 20

158

min, 4ºC), and the culture supernatants filtered through 0.45 µm Durapore HVP nitrocellulose filters. Solid ammonium sulphate was added to the filtrates to give 55% saturation, with stirring (1 h, 4Cº), and the precipitate collected by centrifugation (30000 xg, 15 min, 4ºC). It was then redissolved in a small volume of TCU buffer (150 mM NaCl, 6 M urea, 20 mM Tris-HCl, pH 7) (6). The redissolved pellet was applied to a Sephacryl S-500 column (90 x 2.2 cm), equilibrated and eluted with TCU buffer at 20 ml/h; 6-ml fractions were collected and tested for protein (11), and haemolytic activity. Active fractions were pooled, dialyzed when appropriate and concentrated by ultrafiltration through CX-10 Millipore immersible filters. Dialyzed samples contained less than 5 mM urea.

Chromatofocusing was carried out on a Mono P (HR 5/20) column from Pharmacia, on which a pH gradient (7-3.5) was generated by applying 50 ml of Polybuffer 74-HCl, pH 3.5 (Pharmacia) diluted ten-fold with water. Fractions (1 ml) were collected and pH and haemolytic activity determined, the latter after readjusting pH to 7.

Protein electrophoresis: Dilute samples were precipitated with 10% (w/v) trichloroacetic acid, final concentration (30 min, 4ºC) and centrifuged (30000 xg, 15 min, 4ºC). The precipitates were repeatedly washed with ethanol:ether (1:1, v/v). Concentrated samples, or precipitates from dilute samples, were treated with 2.5% (w/v) SDS, 5% (v/v) β-mercapthoethanol, 0.01 (w/v) bromophenol blue, 100ºC, 5 min. SDS-PAGE was carried out with the Pharmacia Phastsystem (Separation File No. 110). The gels (10-15 polyacrylamide gradient) were developed using the silver or Schiff stains, modified for the Phastsystem procedure.

Chemical analysis and enzyme assays: Proteins were assayed according to Bradford (11) and sugars, by the phenol-sulphuric acid method (12). Lipids were extracted (13) and lipid phosphorus determined after Bartlett (14). Fatty acid analysis was performed by gas-liquid chromatography (15). 2-Keto-3-deoxyoctonate (KDO) was determined colorimetrically by the thiobarbituric acid procedure (16). Dialyzed haemolysin preparations were treated with a variety of hydrolytic enzymes in a medium containing 150 mM NaCl, 20 mM Tris HCl, pH 7. Phospholipase C from Bacillus cereus and egg-white lysozyme were purchased from Boehringer; trypsin from bovine pancreas was supplied by Sigma. Haemolysin preparations were incubated for 10 min at 37ºC (phospholipase C), 15 min at 25ºC (lysozyme), or 15 min at 37ºC (trypsin).

Particle size analysis: Size distributions of particles were determined by photon correlation spectroscopy, also called quasi elastic light scattering (17), using an autosizer IIC (Malvern Instruments, Malvern, U.K.), coupled to a Malvern 7032-N multibit correlator. Steady state light scattering

from haemolysin-containing samples was measured at 90°C using
a RF 540 Shimadzu spectrofluorimeter with both excitation and
emission monochromators adjusted at 500 nm.

Vesicle leakage and lipid mixing: Large unilamellar vesicles
composed of phosphatidylcholine : phosphatidylethanolamine :
cholesterol (PC:PE:CHOL) (2:1:1, mole ratio) were prepared by
extrusion and sized by using 0.1 μm pore-size Nuclepore
membranes as described by Mayer et al. (18). Leakage was
measured by using the ANTS/DPX system according to Ellens et
al. (19). Vesicle lipid mixing was measured by dilution in
the bilayer of the self-quenching probe octadecylrhodamine
(R_{18}) as shown by Hoekstra et al. (20). Experimental details
were as described previously (21).

RESULTS

The data on α-haemolysin purification are summarized in Fig.
1 and Table I. The various purification steps are shown in
the Table; a purification of 251-fold has been achieved. The
apparent yield is much higher than 100%, apparently because
of activation effects. Fig. 1 shows a representative example
of the Sephacryl S-500 column elution; within the limits of
the technique, protein contents and haemolytic activity
present coincident, symmetric peaks, also corresponding to a
peak in A_{280} of the column eluate. (The two large peaks of

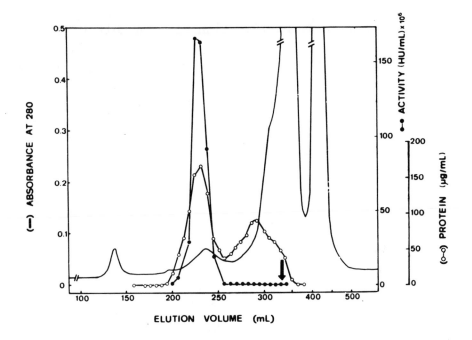

Fig. 1. Elution profile of a representative run in the Sephacryl S-500 column. Continuous line: absorbance at 280 nm; (o)
protein; (●) α-haemolysin activity. After elution of about
300 ml (arrow) flow was increased to 0.6 ml/min.

A_{280} at higher elution volumes correspond to pigments in the

culture medium; they are absent when a chemically defined

defined medium is used instead of LB for growing the bac-

teria). According to silver-stained SDS gel electrophoresis

(not shown), a single polypeptide, corresponding to α-haemo-

lysin activity and with $M_r \approx 110000$ is obtained with a re-

markable degree of purity (7). The isoelectric point of the

α-haemolysin complex has been estimated by chromatofocusing

Table I. Purification of α-haemolysin

PURIFICATION STEP	VOLUME (mL)	PROTEIN (mg/mL) (x 10⁻³)	ACTIVITY (HU/mL)‡ (x 10⁶)	TOTAL ACTIVITY (HU) (x 10⁶)	SPECIFIC ACTIVITY (HU/mg protein) (x 10⁶)	PURIFICATION (FOLD)	YIELD (%)
13000 X g SUPERNATANT	1300	7.7	0.032	42	4.15	1	100
FILTRATE	1300	8.0	0.033	43	4.12	0.99	100
55% A.S.	6.5	1186	147	955	124	30	2273
SEPHACRYL	11.4	160	167	1904	1044	251	4533

‡ The haemolytic unit (HU) is defined under "Materials and Methods".

the Sephacryl S-500 column eluate; a pI = 4.3 is found (Fig. 2).

The chemical composition of our α-haemolysin preparation, as obtained from the gel filtration column, is shown in Table II. The data indicate that it consists essentially of sugar and protein. In view of the possibility that at least part of the carbohydrate component was structurally related to the E. coli lipopolysaccharide, the presence of 2-keto-3-deoxyoctonate (KDO), a typical component of bacterial lipopolysaccharide (22) was investigated, and confirmed. In accordance with this finding, qualitative gas-liquid chromatography indicated that the main fatty acid in our preparation had a retention time compatible with 3-hydroxytetradecanoic

TABLE II. Chemical composition of α- haemolysin
after the last purification step.

Protein	78.0 dry w%
Sugar	19.6 dry wt%
Lipid P	0.85 µg/mg protein
2-keto-3-deoxyoctonate	2.3 µg/mg protein

Average values of two preparations.

acid, a fatty acid known to be associated to lipopolysaccharide (23). When SDS gels are stained by the Schiff method (not shown), only faint bands are stained at the polypeptide level, while large amounts of stained material remain in the stacking gel, suggesting that, although a α-haemolysin may be a glycoprotein, most of the carbohydrate in our preparations is not covalently linked to the protein.

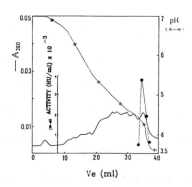

Fig. 2. Chromatofocusing of α-haemolysin purified as in Fig. 1. Continuous line: A280 of the eluate. (∗)pH. (●) Haemolytic activity.

164

The fact that our α-haemolysin preparation is excluded from Sephacryl S-200 columns, whose exclusion limit for globular proteins corresponds to a M ≈ 500 kDa, together with the data of SDS-PAGE electrophoresis and chemical composition, clearly indicate that this preparation consists of large aggregates containing many subunits of sugar and polypeptide. It is difficult to obtain precise indication on the size of aggregate from the Sephacryl S-500 elution data, because of the lack of suitable standards. In addition there were observations, during the purification procedure, that

TABLE III. Average particle size of α-haemolysin aggregates under various conditions

Protein concentration (mg/ml)	6 M Urea	Average diameter (nm)	Polydispersity
0.8	Yes	194	0.51
0.3	Yes	108	0.69
0.8	No	301	0.38
0.3	No	88	0.64

Particle size was estimated by proton correlation spectroscopy. Average values of 3-5 measurements. Polydispersity may vary from 0 (homogeneus size) to 1 (complete heterogeneity).

suggested the possibility of aggregation-dependent changes in activity. Thus, we examined the size of our particles, under various conditions, by a direct method, namely photon correlation spectroscopy. Table III summarizes the data for α-haemolysin at two concentrations, in the presence and absence of urea. Both variables appear to have significant effects on the average particle diameter: increasing protein concentrations leads to an increase in size of the aggregates, while 6 M urea appears to have a concentration-dependent effect, increasing the size of the smaller aggregates and reducing that of the larger ones. Polydispersity, i.e. variability in particle size, is high in all cases, as would be expected for aggregates composed by an indefinite number of subunits.

In view of the above observations, a more detailed study was carried out on the effect of urea and other chaotropic agents on the size and activity of our preparations. For reasons of convenience, absolute determinations of particle size were not carried out, instead steady-state 90ºC ligth-scattering was used to obtain semi-quantitative information on the changes of particle size. The effect of urea is shown in Fig. 3 A. Starting from a dialyzed, urea-free preparation containing 0.1 mg protein/ml, addition of urea up to 6 M produces an increase in haemolytic activity of more than 100-fold, while the scattering falls gradually to about one-half the original value. Guanidinium chloride, also known as a

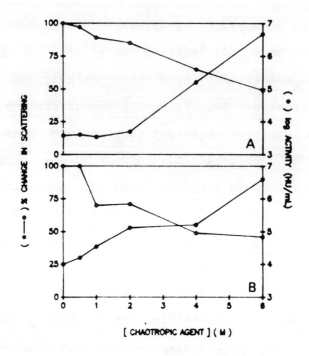

Fig. 3. The effect of chaotropic agents on the size and activity of α-haemolysin aggregates. (A) urea; (B) guanidinium chloride. (o) Light-scattering from the protein suspension; 100% is the scattering in the absence of chaotropic agents. (●)α-haemolysin activity.

chaotropic agent, behaves very much like urea (Fig. 3 B), increasing haemolytic activity and decreasing scattering, the latter suggesting a decrease in aggregate size.

The tendency of α-haemolysin to aggregate in aqueous media points towards the possibility of hydrophobic forces mediating the interaction. In order to test this hypothesis, a dialyzed α-haemolysin preparation (0.13 mg protein/ml) was treated with the non-ionic surfactant Triton X-100, which is known to effectively disperse aggregates held together by hydrophobic forces, while preserving protein activity. The results in Fig. 4 suggest that 0.1% (v/v) Triton X-100 (about 1.6 mM) is as effective as 6 M urea in decreasing aggregate size, thus supporting the idea that hydrophobic forces are important in α-haemolysin aggregation.

Another non-ionic surfactant, Tween 80, had also similar effects on our protein (not shown). The effect of these detergents on α-haemolysin activity, also depicted in Fig. 4

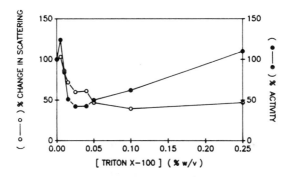

Fig. 4. The effect of Triton X-100 on the size and activity of α-haemolysin aggregates. (o) Light-scattering from the protein suspension; 100% is the scattering in the absence of detergent. (●)α-haemolysin activity.

for Triton X-100, is complex, and the kind of activation produced by urea or guanidinium chloride is not observed here. It should be noted that, for all data on haemolytic activity of surfactant-treated α-haemolysin, the haemolytic effects of the detergents themselves were checked, and found to be negligible, at the α-haemolysin dilutions used in the assays, in accordance with previous studies (24).

168

Even in the absence of chaotropic agents of surfactants, the specific activity of α-haemolysin changes with concentration, as seen in Fig. 5. Samples obtained by dilution of the Sephacryl S-500 eluate show that activity increases first, and then decreases with concentration.

As a further test of the nature of the α-haemolysin aggregates, both the purified preparation and the crude filtrate were treated with a series of hydrolytic enzymes,

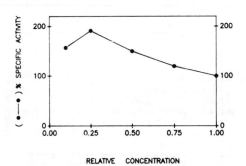

Fig. 5. The effect of dilution on the specific activity of α-haemolysin. Dialyzed samples containing about 100 µg protein/ml were diluted with buffer and their activities tested. Reactive concentration 1.0 corresponds to the undiluted sample. Values are expressed as percentages, 100% being the specific activity of the undiluted sample (≈ 69000 HU/mg protein). Average values of three experiments.

and their effect on haemolytic activity tested. Phospholipase C, at concentrations up to 150 units/ml had no effect whatsoever, and the same was true of lysozyme, at concentrations up to 500 units/ml. It should be stressed that those enzymes were tested with appropriate substrates under the same conditions of our experiment, and the expected activities were found. On the other hand, the pure enzymes, after being

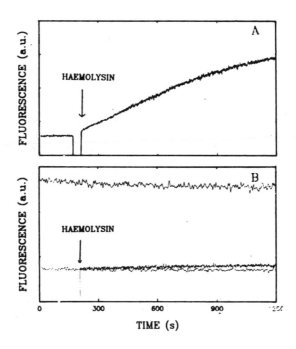

Fig. 6. The interaction of α-haemolysin with phospholipid vesicles. (A) Vesicle leakage, as measured by the ANTS/DPX system. (B) Lipid mixing assay, followed as the dequenching of octadecylrhodamine; the lower superimposed traces correspond to the liposome suspension in the presence and absence of protein, while the upper trace was recorded in the presence of 1mM Triton X-100 (corresponding to 100% lipid mixing). Phospholipid concentration: 0.1 mM; α-haemolysin concentration: 0.1 μM.

diluted under the same conditions as α-haemolysin, had no detectable haemolytic effect by themselves. Trypsin, however, was highly deleterious for α-haemolysin activity: incubation with 0.25% (w/w) for 15 min at 37ºC completely destroyed the activity.

The interaction of α-haemolysin with lipid bilayers was tested using large unilamellar vesicles composed of PC/PE/CHOL (2:1:1 mole ratio). When vesicles containing both ANTS and DPX are treated with α-haemolysin, an increase in fluorescence is observed (Fig. 6A) which is interpreted as release of vesicle contents (19,21). However, α-haemolysin does not promote any change in fluorescence when R18-containing and R$_{18}$-free liposomes are mixed in the presence of the protein (Fig. 6B) indicating that lipid mixing does not occur under these conditions (20,21).

DISCUSSION

The available data on E. coli α-haemolysin composition and properties are in some cases confusing, and even contradictory. Our work intends to clarify at least some of these aspects, particularly in what refers to (a) purification and chemical composition, (b) physical properties, and (c) structure-function relationships.

Our α-haemolysin preparation is among the purest described up to now, judging from the silver-stained gels (7). Only Wagner et al. (25) have published comparable gels. This demonstrates the important point that a single 110 kDa polypeptide makes up for the protein moiety of haemolysin, as suggested by previous studies (6,26) but not by the data of Bohach and Snyder (23). With respect to the purification method by González-Carreró et al. (6), our procedure has the advantage of using a more efficient plasmid; α-haemolysin titres in supernatants from cultures containing the pSU 157 plasmid are of about 800 HU/ml, while, with the pSU 124 plasmid, the corresponding values are around 20000 HU/ml. Our preparation has a considerably higher specific activity (about 10^9 HU/mg protein vs. 5.4 x 10^6 HU/mg protein in their paper), and the protein concentration in the final eluate is four times higher in our case. Also, the use of a Sephacryl S-500 column, in which α-haemolysin aggregates penetrate, instead of the one used by González-Carreró et al. (6), from which the aggregates were excluded, may improve the purity of our preparations. From the point of view of chemical composition, the heterogeneity of published data is remarkable. The matter has been reviewed by Cavalieri et al. (1) and by Bhakdi et al. (2). Cavalieri et al. (1) based on previous studies (27,28) suggest that carbohydrate is absent from α-haemolysin preparations when cells have been grown in complex media, but there are no reliable data in the literature to confirm or deny this assertion. Our preparation (from a complex medium) contains ≈ 20 dry wt% sugar although, as discussed in the Results section, most if not all of the carbohydrate is noncovalently linked to 110 kDa polypeptide. The presence of 2-keto-3-deoxyoctonate and 3-hydroxytetradecanoic acid suggests almost with certainty the occurrence of lipopolysaccharide in our preparation (22,29); this had been already found by Bohach and Snyder (23) in

172

their sugar-rich preparation, and their observation cautiously recorded by Bhakdi et al (2). The presence of lipopolysaccharide may also account for the amounts of lipid phosphorus present in our preparation (Table II) (29). It should be noted, in this respect, that phospholipase C treatments do not modify α-haemolysin activity in our preparations. Wagner et al. (25) find α-haemolysin to be highly sensitive to phospholipase C; in spite of detailed experimentation, we have been unable to reproduce these results with our preparation, nor can we suggest a hypothesis to explain the difference. Moreover, the heterogeneity of the preparations suggests that, in the largely unknown process of protein export, the polypeptide may bind, in a non-stoichiometric way, some portions or components of the bacterial envelopes.

There have been various suggestions that α-haemolysin is multimeric, or occurs in the form of aggregates (1,6,30). Bohach and Snyder (23) propose that the aggregate size is greater than 60 nm, and that considerable heterogeneity in size occurs. Our photon correlation spectroscopic measurements provide support to both assumptions (Table III), with the additional significant observation that the average size changes with the presence or absence of 6 M urea. Another important physical property of proteins is their isoelectric point; the value found for our preparation (pI = 4.3) is comparable to previous measurements (28,31). Bohach and

Snyder (23) found a lower value, around 3.6, but the difference may be due to the very high carbohydrate contents of their preparations.

Various authors have pointed out that, since α-haemolysin activity is destroyed by trypsin, the protein moiety of the aggregates must be essential for haemolytic activity (1). However, this paper describes for the first time the inhibitory action of trypsin on a α-haemolysin preparation whose protein component is only the 110 kDa polypeptide, thus relating directly this particular polypeptide to α-haemolysin activity. The role of non-protein components in haemolytic activity, suggested by several workers (23,25) is not sufficiently documented as yet. The relationship between activity and degree of aggregation appears to be rather complex. Such possibility was mentioned already by González-Carreró et al. (6) but specific experiments on this matter have not been presented until now. Our data in Figs. 3 and 5 indicate that factors (chaotropic agents, dilution) favouring disaggregation, also increase the specific activity of α-haemolysin. However Fig. 5 (data at high dilution) suggests that the concentration effect may be more complex, increasing activity within certain limits and decreasing it beyond a given point. These observations explain the large increases of total activity found during α-haemolysin purification by us and other authors, and also the data of Wagner et al. (3) showing a lack of correlation between

total extracellular α-haemolysin and haemolytic activity of
the medium. Also, our conclusion that disaggregating factors
increase specific activity is in accord with the suggestion
by Bhakdi et al. (2,32) that α-haemolysin binds erythrocytes
in the monomer form. In connection with the stabilizing and
activating properties of urea, Roepe and Kaback (29) have
recently reported the interesting observation that overpro-
duced lac permease from E. coli, but not the membrane-
inserted form, can be solubilized by 5.0M urea in the absence of detergents; when urea is removed, the protein
aggregates slowly. The water-soluble permease associates
with E. coli phospholipids, just as α-haemolysin interacts
with liposomes (Fig.6). Structural studies of these and
other amphitropic proteins, in the various environments in
which they appear to remain physiologically active, will
certainly reveal interesting aspects of potential biological
and technological value.

Acknowledgments: This work was supported in part by grants
No. PB 88-0301 and BIO 88-0407-C02-01 from DGICYT.

The authors are grateful to J. Hernández Borrell and J.
Sterlich for the photon correlation measurements; A. Juárez
and J. Tomás for their help with the KDO determinations, and
J.C. Zabala for the gift of the pSU 124.

REFERENCES

1. Cavalieri, S.J., Bohach, G.A. and Snyder, I.S. (1984) Microbiol. Rev. 48, 326-343.
2. Bhakdi, S., Mackman, N., Menestrina, G., Gray, L., Hugo, F., Saeger, W. and Holland, I.B. (1988) Eur. J. Epidemiol. 4, 135-143.
3. Mackman, N., Baker, K., Gray, L., Haigh, R., Nicaud, J.M. and Holland, I.B. (1987) EMBO J. 6, 2835-2841.
4. Wagner, W., Vogel, M. and Goebel, W. (1983) J. Bacteriol. 154, 200-210.
5. Nicaud, J.M., Mackman, N., Gray, L. and Holland, I.B. (1985) Mol. Gen. Genet. 199, 111-116.
6. González-Carreró, M.I., Zabala, J.C., de la Cruz, F. and Ortiz, J.M. (1985) Mol. Gen. Genet. 199, 106-110.
7. Ostolaza, H. Bartolomé, Serra, J.L., de la Cruz, F. and Goñi, F.M. (1991) FEBS Lett. 280, 195-198.
8. Zabala, J.C., García-Lobo, J.M., Díaz-Aroca, E., de la Cruz, F. and Ortiz, J.M. (1984) Mol. Gen. Genet. 197, 90-97.
9. Snyder, I.S. and Koch, N.A. (1966) J. Bacteriol. 91, 763-767.
10. Miller, I.H. (1972) "Experiments in Molecular Genetics". Cold Spring Harbor Laboratory, Cold Spring Harbor, N.Y.
11. Bradford, M.M. (1976) Anal. Biochem. 72, 248-254.
12. Dubois, M., Gilles, K.A., Hamilton, J.K., Rebers, P.A. and Smith, F. (1956) Anal. Chem. 28, 350-356.
13. Santiago, E., Mule, S., Redman, M., Hokin, M.R. and Hokin, L.E. (1964) Biochim. Biophys. Acta 84, 550-562.
14. Bartlett, G.R. (1959) J. Biol. Chem. 234, 466-468.
15. Regúlez, P., Pontón, J., Domínguez, J.B., Goñi, F.M. and Uruburu, F. (1980) Can. J. Microbiol. 26, 1428-1437.
16. Weissbach, A. and Hurwitz, J. (1959) J. Biol. Chem. 234, 705-709.
17. McConnell, M.L. (1981) Anal. Chem. 53, 1007 A-1018 A.
18. Mayer, L.D., Hope, M.J., and Cullis, P.R. (1986) Biochim. Biophys. Acta 858, 161-168.
19. Ellens, H., Bentz, J., and Szoka, F.C. (1985) Biochemistry 24, 3099-3106.
20. Hoekstra, D., De Boer, T., Klappe, K., and Wilschut, J. (1984) Biochemistry 23, 5675-5681.
21. Nieva, J.L., Goñi, F.M., and Alonso, A. (1989) Biochemistry 28, 7364-7367.
22. Osborn, M.J. (1963) Proc. Nat. Acad. Sci. U.S.A. 50, 499-506
23. Bohach, G.A. and Snyder, I.S. (1985) J. Bacteriol. 164, 1071-1080.
24. Wagner, W., Kuhn, M. and Goebel, W. (1988) Biol. Chem. Hoppe-Seyler 369, 39-46.
25. Mackman, N. and Holland, I.B. (1984) Mol. Gen. Genet. 193, 312-315.

26. Williams, P.H. (1979) FEMS Microbiol. Lett. 5, 21-24.
27. Bhakdi, S., Mackman, N., Nicaud, J.M. and Holland, I.B. (1986) Infect. Immun. 52, 63-69.
28. Rennie, R.P. and Arbuthnott, J.P. (1974) J. Med. Microbiol. 7, 179-188.
29. Nikaido, H. and Varaa, M. (1987). In:"E. coli and S. typhimurium cellular and molecular biology". American Society of Microbiology, Washington, D.C., pp. 7-22.
30. Williams, P.H. (1979) FEMS Microbiol. Lett. 5, 21-24.
31. Short, E.C. and Kurtz, H.J. (1971) Infect Immun. 3, 678-687.
32. Bhakdi, S., Mackman, N., Nicaud, J.M. and Holland, I.B. (1986) Infect. Immun. 52, 63-69.
33. Roepe, P.D. and Kaback, H.R. (1989) Proc. Natl. Acad. Sci. USA 86, 6087-6091.

Progress in Membrane Biotechnology
Gomez-Fernandez/Chapman/Packer (eds.)
© 1991 Birkhäuser Verlag Basel/Switzerland

ON THE MECHANISM OF PHOSPHOLIPASE C-INDUCED FUSION OF PURE

LIPID MEMBRANES

José-Luis Nieva and Alicia Alonso

Department of Biochemistry, University of the Basque Country,
P.O. Box 644, 48080 Bilbao, Spain

SUMMARY: Biochemical, electron microscopic and ^{31}P-NMR spectroscopic techniques have been applied to the study of fusion
of large unilamellar vesicles, consisting of phospholipid and
cholesterol, in the presence of diacylglycerol, either externally added or generated in situ by phospholipase C. The
processes of diacylglycerol production by the enzyme, vesicle
aggregation and vesicle fusion have been dissected and
studied separately. Fusion appears to proceed through the
(transient) formation of non-lamellar lipid structures.

In the last decade, the crucial role of phospholipid

metabolism in cell activation has been clearly established

(Pelech and Vance, 1989). In particular, secretory processes

proceed as a result of phospholipid cleavage and diacyl-

glycerol (DG) production in the membrane. Also, studies using

model membranes have shown that DG are potent modulators of

membrane structure; Das and Rand (1986) indicated that the

local DG production by phospholipase C may be directly in-

volved in the membrane fusion process.

The combination of fluorescence spectroscopy and lipo-some technology (Wilschut and Papahadjopoulos, 1979; Ellens et al., 1985) provides the possibility of assessing the fuso-genic abilities of different substances in a model system, and studying their mechanism of action. Wilschut et al. (1980, 1985) were able to distinguish between membrane ag-gregation or apposition, and fusion or destabilization of the membranes in apposition, the latter step including mixing of aqueous vesicle contents. A different approach includes studies with liposomes containing lipids, such as PE, that exhibit thermotropic transitions to non-lamellar phases (Gag-né et al., 1985; Ellens et al., 1986). Fusion appears to occur preferentially at or near the transition temperature; it has been suggested that non-lamellar intermediates would be formed, which in turn would allow the mixing of aqueous contents (Siegel et al., 1989; Burger and Verkleij, 1990). In this case, the aggregation and fusion processes cannot be separated, since the conditions allowing aggregation also produce the mixing of lipids and of vesicle contents (Allen et al., 1990).

In our previous paper (Nieva et al., 1989), it was demonstrated that an enzyme, phospholipase C from Bacillus cereus, was able to induce aggregation, and then fusion, of liposomes. In the present study a mechanism is proposed according to which the local production of diglycerides is responsible, first, for the massive liposomal aggregation, and then, for the membrane destabilization that allows the mixing of vesicle contents. In addition, the existence of two possible destabilization pathways is shown, one requiring high diglyceride (DG) concentrations, probably involved in the disorganization of membrane structures inside the aggregates, and a second one, sensitive to membrane composition and assay temperatures, that proceeds at low DG concentrations in the membrane, and is saturated and inhibited by high DG. It is suggested that the latter pathway may be modulated by intermediates of a defined non-lamellar structure.

MATERIALS AND METHODS

Phospholipase C (EC 3.1.4.1) from Bacillus cereus was supplied by Boehringer-Mannheim. Egg phosphatidylcholine (PC) and phosphatidylethanolamine (PE) were from Lipid Products (South Nutfield, England); cholesterol (CHOL) was from Sigma (St. Louis, MO). 1,3,6-trisulphonate-8-aminonaphthalene (ANTS) and p-xilenebis (pyridinium bromide) (DPX) were purchased from Molecular Probes (Eugene, OR).

Substrate preparation: Large unilamellar vesicles (LUV) prepared by the extrusion method of Mayer et al. (1986) were used as the phospholipase C substrate. The lipid composition of these liposomes was PC:PE:CHOL (2:1:1 mol ratio). The lipid suspension was extruded through Nuclepore filters, 0.1μm, pore diameter. Freeze fracture electron microscopic measurements indicated an apparent average diameter of 114 nm.

Fusion assays: Mixing of aqueous vesicle contents, as well as vesicle leakage, were estimated using the ANTS/DPX fluorescent probe system described by Ellens et al. (1985). Three liposome preparations were used, loaded with (a) 50 mM ANTS, 90 mM NaCl, 10 mM CaCl$_2$, 10 mM HEPES, pH 7.0; (b) 180 mM DPX, 10 mM CaCl$_2$, 10 mM HEPES, pH 7.0; or (c) 25 mM ANTS, 90 mM DPX, 45 mM NaCl, 10 mM CaCl$_2$, 10 mM HEPES, pH 7.0. When appropriate, 2 mM CaCl$_2$ instead of 10 mM CaCl$_2$ was used. Non-encapsulated material was removed from the vesicles using a Sephadex G75 column, with 10 mM HEPES, 200 mM NaCl, 10 mM CaCl$_2$ (or 2 mM CaCl$_2$), pH 7.0 as the elution buffer. This buffer was also used in all the fusion and enzyme assays. The osmolalities of all solutions were measured in a cryoscopic osmometer (Osmomat 030, Gonotec, Berlin, Germany) and adjusted to 0.4 osmol/kg by adding NaCl.

Fluorescence scales were calibrated for fusion and leakage assays as described previously (Nieva et al., 1989). Assays were performed in thermostatted cuvettes with constant stirring, in a LS-50 Perkin Elmer spectrofluorometer. Excitation light was adjusted at 355 nm, and emission, at 520 nm. An interference filter (450 nm) was used to avoid scattered excitation light. Liposome aggregation was estimated by fixing the excitation and emission monochromators at 520 nm.

Enzyme assays were carried out as described by Nieva et al. (1989).

Electron microscopy: Experiments were performed in the presence of 30% (v/v) glycerol. Small aliquots were transferred to gold platelets and frozen from room temperature by dipping the samples either into a mixture of solid and liquid nitrogen or into liquid propane cooled to its melting point. Freeze fracture and replication were carried out following standard procedures. The replicas were stripped off on water and cleaned with commercial bleach and distilled water. The replicas were examined with a Philips CM10 electron microscope at 100 kV.

^{31}P-NMR spectroscopy: ^{31}P-NMR spectra were recorded in a Varian KM 360 spectrometer, equipped with a thermostatted sample holder. Lipid concentration was 0.2 M. The following spectral parameters were used: pulse width 10 µs, sweep width 16 kHz, time interval between pulses 3 s. About 1000 spectra were routinely accumulated at each temperature for a given sample. Line broadening was 80 Hz.

RESULTS AND DISCUSSION

Ultrastructural studies: The conventional freeze-fracture method requires that experiments are performed in the presence of 30% (v/v) glycerol. Independent measurement showed that glycerol slowed down the kinetics of the fusion process, though it hardly affected phospholipase C activity (data not

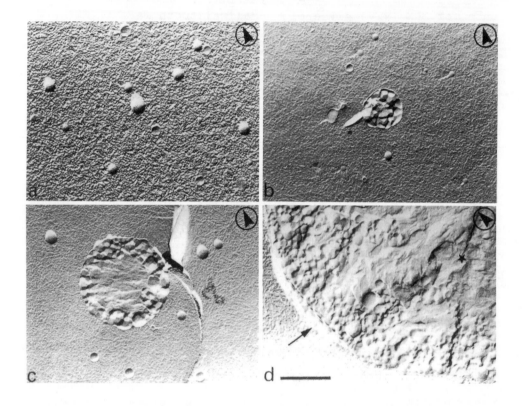

Fig. 1. A freeze-fracture electron microscopy study of phospholipase C-induced aggregation and fusion of large unilamellar vesicles. Aliquots from the sample were frozen 0 s (a), 30 s (b), 90 s (c), and 300 s (d) after enzyme addition. The aggregates appear to be surrounded by a continuous membrane (arrow in (d), inner fracture face), and the aggregate core presents an amorphous appearance (star in (d). Lipid concentration: 10 mM; enzyme concentration: 400 U/ml. All pictures printed at the same magnification; bar = 500 nm. The arrow head indicates the direction of the Pt/C shadowing. (Pictures supplied by K. Burger and A. Verkleij).

shown). Samples were taken from a suspension of LUV and frozen, before, and 30,90 or 300 s after adding phospholipase C. At these stages DG contents, arising from phospholipid cleavage, were about 0,15 and 60 mol % (of total lipid) respectively. The corresponding freeze fracture results are shown in Fig. 1; vesicle aggregates are seen to form and to increase in size with time. The outer membrane of these aggregates appears to be continuous (Fig. 1b-d) and this enveloping membrane must have resulted from fusion processes. At an early timepoint (30 s, Fig. 1b) other signs of bilayer destabilization are not observed, though it should be noted that the vesicular compartments of the aggregate have a peculiar polygonal appearance. At longer incubation times (90 and 300 s, Fig. 1c and d respectively) the vesicular compartments decrease in size and, in addition, the aggregate core progressively looses its bilayer structure becoming almost amorphous in appearance (marked by star in Fig. 1d). The freeze fracture images show no sign of an intermembrane space separating neighbouring vesicular compartments; instead the images suggest that neighbouring compartments share one bilayer.

The images presented in Fig. 1 suggest that aggregates grow peripherally via fast aggregation and non-leaky fusion (see Nieva et al., 1989), for which only small amounts of DG are required, while a slower process of extensive phospholipid hydrolysis could be responsible for the changes observed in the core structure; the amorphous core may well contain DG and CHOL almost exclusively. The fact that aggregates always appear to be surrounded by a continuous membrane may explain the absence of (detectable) vesicle leakage even when substantial phospholipid hydrolysis has occurred. This finding constitutes an important warning to those relying on biochemical (dilution) assays to determine whether vesicle fusion does or does not involve leakage of internal contents: these assays will fail to detect leakage if it occurs within a ('sealed') aggregate of vesicles.

Biochemical studies: Phospholipase C from Bacillus cereus is highly effective in promoting the fusion of large unilamellar liposomes composed of PC:PE:CHOL (2:1:1). As shown in Fig. 2, the fusion process may be modified by substituting for DG some of the phospholipids in the initial liposome formulation. At 37ºC, when no DG is present in the substrate (Fig. 2A), fusion starts about 10 s after enzyme addition, when the amount of generated DG is about 4% of the total lipid. The reaction proceeds for another 20 s (up to ≈ 15% DG), before end-product inhibition becomes apparent; still, although at declining rates, the process continues until virtually 100% mixing occurs, but without any significant vesicle leakage. When the initial liposome composition is modified, so that 5% of the lipids is substituted by DG, the fusion process becomes faster. The lag time after enzyme addition is shorter;

10 s after the addition the total DG content is 12%, and the
reaction rate is maximal. Saturation occurs at ≈ 20 s after
enzyme addition, when DG makes up 20% of the total lipids. In

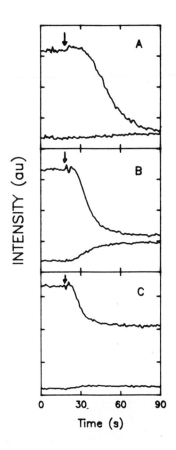

Fig. 2. Time-course of the phospholipase C-induced fusion of
PC:PE:CHOL (2:1:1) vesicles, studied with the ANTS and DPX
fluorescence probes. In each panel, the upper trace corresponds to the contents mixing assay, and the lower trace, to
the leakage assay, The enzyme (1.6 U/mL) is added at t = 20 s
(arrow). (A) Liposomes not containing DG prior to enzyme
addition. (B) Liposomes containing initially 5% DG. (C) Liposomes containing initially 10% DG. Temperature was 37ºC in
all cases.

this case, vesicle contents are partially released, in a process that occurs simultaneously with contents mixing. When the substrate liposomes contain 10% of the lipid substituted by DG, fusion starts as soon as phospholipase C is added, and the process is rapidly saturated after 10 s, when the total amount of DG is $\approx$ 20%. These results appear to indicate that, at 37ºC, the enzyme catalyzes efficiently the fusion of liposomes containing PC:PE:CHOL (2:1:1) plus DG, when the proportion of the latter is in the range 5-20% of the total lipid. At higher DG proportions, the process becomes saturated by this end-product. The fact that, in liposomes containing originally DG, some degree of leakage occurs, may indicate that the membrane destabilization event that is at the origin of vesicle fusion may also lead to release of contents when it occurs in the periphery of an aggregate.

The observed fusion process is abolished when lyso PC is incorporated into the liposomes at a 10% mol ratio, or when Ca^{++} concentration is reduced from 10 to 2 mM (data not shown). In these samples fusion occurs only at DG concentrations higher than 30%. Fast fusion is also absent in pure PC liposomes, where a slow process of contents mixing occurs in the 40-60% interval of DG concentrations, until saturation takes place.

A detailed kinetic analysis considering the various liposome preparations as a function of temperature shows that fusion varies with temperature in a different way than enzyme activity, or vesicle aggregation (Fig. 3). A good correlation

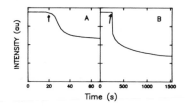

Fig. 3. Time-course of the phospholipase C-induced fusion of PC:PE:CHO (2:1:1) liposomes, containing initially 10% DG. (A) Fast process. (B) Slow process.

is observed, at all temperatures, between aggregation and enzyme activity, except for the liposomes containing originally DG. In the other cases, both phenomena appear to be activated at ≈ 45ºC. For DG-containing liposomes, the enzyme activity appears to be higher at all temperatures, but the temperature dependence is otherwise similar to the other substrates. However, aggregation is much more efficient in the presence of DG than in its absence, particularly at low temperatures, suggesting that DG confers the vesicles some degree of adherence.

The temperature dependence of vesicle fusion (as mixing of contents) shows maxima whose absolute value is higher, and occurs at lower temperatures, the higher the initial DG contents. The presence of lyso PC, or low Ca^{++} concentrations, specifically inhibit this process. The increased

fusion in the DG-containing samples could be due to increased
vesicle adherence. In contrast with aggregation, fusion does
not increase indefinitely with temperature, and each liposome
composition exhibits a peculiar pattern as a function of
temperature. These data suggest a certain independence between the fusion and aggregation processes.

The independent behaviour of those two processes is also
confirmed by the fact that, as shown in Fig. 4, the lapse of
time required to reach a maximum of fusogenic activity is
longer than the corresponding time for attaining the maximum
aggregation rate. In samples containing originally DG, the
time lag between the two phenomena is sufficient to ensure
that fusion is not limited by aggregation. Vesicle leakage
has also been studied as a function of temperature, for
liposomes containing initially 5% DG. The data show that both
fusion and leakage display parallel patterns as a function of
temperature, suggesting that both processes are closely related and may share a common mechanism.

According to our results the enzyme activity, and not
the mere presence of DG in the bilayer, is responsible for
the fusion process. Liposomes containing 10% DG were stable
even at 80ºC in the absence of the enzyme. This may suggest
that the localized action of the enzyme on the outer monolayer is crucial for the initial aggregation steps. Das and
Rand (1986) demonstrated that the inter-bilayer equilibrium
was not modified in the presence of added DG, which cor-

relates well with the lack of effect observed in the absence

of enzyme. We have also shown that the fusion process can be

decomposed into two components; one is fast and very ef-

fective, and occurs when the DG concentration in the bilayer

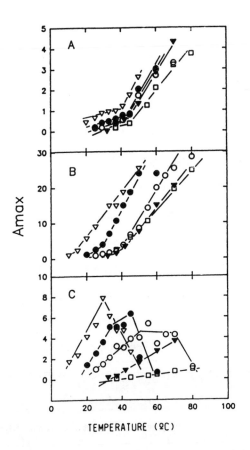

Fig. 4. Maximal activities recorded as a function of temperature after addition of 1.6 U/mL phospholipase C. (A) Phospho-
hydrolase activities. (B) Vesicle aggregation rates. (C)
Fusion rates. (○) PC:PE:CHOL (2:1:1). (●) Id. + 5% DG. (▽)
Id. + 10% DG. (▼) ID. + 10% lyso PC. (□) 2 mM instead of 10
mM Ca^{++}.

190

is between 5 and 20%. This fast component becomes saturated at higher DG concentrations, and is also inhibited by temperature, lysophosphatidylcholine (lyso PC) and by low Ca^{++} concentrations. The fast fusion component is occasionally accompanied by the release of vesicle contents. The second component is much slower, and is easily detected in samples in which the fast process is either saturated or inhibited, it proceeds even with high DG concentrations, and could be reflecting the gradual loss of the lamellar structure within the aggregates.

The temperature dependence of the fast fusion process is different from that of aggregation and enzyme activity. At low temperatures, fusion proceeds at a slow rate, but it is very efficient (saturation values near 100%); when temperature is increased, the rate increases, but the saturation point decreases. Finally, at the highest temperatures tested, both values are decreased. This inhibitory effect, at temperatures at which both enzyme activity and vesicle aggregation are enhanced, may reflect the existence of a structural intermediate whose formation would be influenced by temperature and by DG concentration. The fact that lyso PC also inhibits the fast fusion process suggests that the intermediate may have a non-lamellar structure, as in the case of temperature-dependent fusion in systems containing egg PE or methyl DOPE (Ellens et al., 1986 a, b). The existence of such an intermediate would be in accordance with the destabilizing properties of DG, described for a variety of systems (Das and Rand, 1986), and with the rich thermotropic polymorphism exhibited by PC:PE:CHOL mixtures (Cullis and de Kruijff, 1978). The effect of Ca^{++} as a specific modulator of the fusion process is difficult to explain at the moment, and deserves further experimentation. Our results also open the way for independently exploring vesicle aggregation and fusion as a result of phospholipase C activity.

NMR spectroscopic studies: A series of ^{31}P-NMR spectra of aqueous dispersions of PC:PE:CHOL (2:1:1 mole ratio) have been recorded at various temperatures, in the absence and presence of 5% diacylglycerol (DG). In the absence of DG (Fig. 5) the spectral lineshape typical of lamellar phases is found throughout the 20-70ºC temperature interval: a peak at high field, with a shoulder at low field (Seelig, 1978). However, in the presence of 5% DG, an isotropic signal appears (Fig. 6), coexisting with the lamellar phase around 60ºC, and becoming predominant at higher temperatures. The isotropic signal remains unchanged when the sample is cooled down to 20ºC (spectrum not shown).

Our NMR data show that even a relatively small proportion of DG leads to the destabilization of lamellar systems formed by PC:PE:CHOL (2:1:1). Such a destabilizing effect of DG has been described in other systems, where isotropic signals have also been observed in NMR spectra (Tilcock et al., 1982; Gagné et al., 1985; Gruner et al., 1988; Shyam-

sunder et al., 1988). It has also been suggested, from elec-

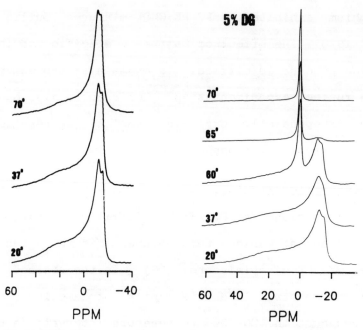

Fig. 5. ³¹P-NMR spectra of an aqueous dispersion of PC:PE:CHOL (2:1:1) as a function of temperature.

Fig. 6. ³¹P-NMR spectra of an aqueous dispersion of PC:PE:CHOL (2:1:1) containing 5% DG, as a function of temperature.

tron microscopy data (Frederik et al., 1991), that structures

giving rise to isotropic signals could correspond to lipids

partially organized in bicontinuous cubic phases, or their

precursors. The hypothesis that low proportions of DG may

induce the formation of a bicontinuous cubic phase is at-

tractive because, in view of the macroscopic description of

such phases (Mariani et al., 1988), it could be speculated

that phospholipase C induces vesicle fusion (mixing of contents) when a small, localized amount of DG destabilizes the

bilayer arrangement and induces the formation of lipid chan-

nels between liposomes in apposition.

Acknowledgments: This work was supported in part with funds from the Basque Government (042.310-0001/88), DGICY (PB 88/0301) and EEC-Science (SC1-0195-C).

REFERENCES

Allen, T.M., Hong, K., and Papahadjopoulos, D. (199) Biochemistry 29, 2976-2985.
Burger, K.N.J., and Verkleij, A.J. (1990) Experient , 46, 631-644.
Cullis, P.R., and De Kruijff, B. (1978) Biochim. Biophys. Acta 507, 207-218.
Das, S., and Rand, R.P. (1986) Biochemistry 25, 2882-2889
Ellens, H., Bentz, J., and Szoka, F.C. (1985) Biochemistry 24, 3099-3106.
Ellens, H., Bentz, J. and Szoka, F.C. (1986a) Biochemistry 25, 285-294.
Ellens, H., Bentz, J., and Szoka, F.C. (1986b) Biochemistry 25, 4141-4147.
Frederik, P.M., Burger, K.N.J., Stuart, M.C.A., and Verkleij, A.J. (1991) Biochim. Biophys. Acta. In the press.
Gagné, J., Stamatos, L., Diacovo, T., Hui, S.W., Yeagle, P.L., and Silvius, J.R. (1985) Biochemistry 24, 4400-4408.
Gruner, S.M., Tate, M.W., Kick, G.L., So, P.T.C., Turner, D.C., Keane, D.T., Tilcock, C.P.S., and Cullis, P.R. (1988) Biochemistry 27, 2853-2866.
Mariani, P., Luzzati, V., and Delacroix, H. (1988) J. Mol. Biol., 204, 165-189.
Mayer, L.D., Hope, M.J., and Cullis, P.R. (1986) Biochim. Biophys. Acta 858, 161-168.
Nieva, J.L., Goñi, F.M., and Alonso, A. (1989) Biochemistry 28, 7364-7367.
Pelech, S.L., and Vance, D.E. (1989) Trends Biochem. Sci., 14, 289-300.
Seelig, J. (1978) Biochim. Biophys. Acta 545, 105-140.
Shyamsunder, E., Gruner, S.M., Tate, M.W., Turner, D.C., So, T.C., and Tilcock, C.P.S. (1988) Biochemistry 27, 2332-2336.
Siegel, D.P., Banschbach, J., Alford, D., Ellens, H., Lis,

194

J., Quinn, P.J., Yeagle, P.L., and Bentz, J. (1989a) Biochemistry 28, 3703-3709.

Siegel, D.P., Burns, J.L., Chestunt, M.H., and Talmon, Y. (1989b) Biophys. J., 56, 161-169.

Tilcock, C.P.S., Bally, M.B., Farren, S.B., and Cullis, P.R. (1982) Biochemistry 21, 4596-4601.

Wilschut, J., and Papahadjopoulos, D. (1979) Nature 281, 690-692.

Wilschut, J., Duzgünes, N., Fraley, R., and Papahadjopoulos, D. (1980) Biochemistry 19, 6011-6021.

Wilschut, J., Nir, S., Sholma, J., and Hoekstra, D. (1985) Biochemistry 24, 4630-4636.

Progress in Membrane Biotechnology
Gomez-Fernandez/Chapman/Packer (eds.)
© 1991 Birkhäuser Verlag Basel/Switzerland

LIPOSOMES AS VEHICLES FOR VACCINES: INTRACELLULAR FATE OF LIPOSOMAL ANTIGEN IN MACROPHAGES

Carl R. Alving*, Nabila M. Wassef*, Jitendra N. Verma*, Roberta L. Richards*, Carter T. Atkinson†, and Masamichi Aikawa†

*Department of Membrane Biochemistry, Walter Reed Army Institute of Research, Washington, DC 20307-5100, U.S.A., and †Institute of Pathology, School of Medicine, Case Western Reserve University, Cleveland, OH, U.S.A.

SUMMARY: Liposomes are highly effective as carriers of antigens and adjuvants for induction of humoral immunity. Not only are high levels of serum antibodies obtained, but unique conformational specificities of antibodies against synthetic peptides can be achieved. It is presumed that the macrophage serves as an antigen presenting cell for liposomal antigen, and the intracellular destinations of liposomes and liposomal antigens in macrophages are therefore of interest. The intracellular fate of phagocytosed liposomal malaria antigen was examined by ELISA and by immunogold electron microscopy, and cytoplasmic delivery of liposomal antigen was observed. It is proposed that transfer of phagocytosed liposomal antigen from intracellular vacuoles through the cytoplasm to the macrophage surface represents a unique route by which liposomal antigens may be processed by macrophages.

IMMUNE RESPONSE TO LIPOSOMAL ANTIGENS

Recombinant proteins: The availability of genetically engineered synthetic peptides and proteins has resulted in a plentiful supply of safe and well-defined candidate antigens for vaccines. Unfortunately, in many instances greatly simplified antigens are

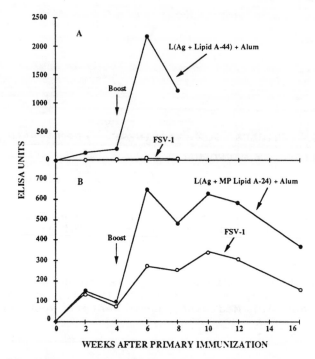

Figure 1.   Time course of immune responses induced by the FSV-1 vaccine or liposomal vaccines in monkeys.   Each point represents the mean value for four monkeys after subtraction of the preimmunization value.   Each monkey was immunized at 0, 4, and 8 weeks with FSV-1 or liposome vaccine, each containing either (A) 30 μg or (B) 80 μg of R32tet$_{32}$ antigen.   ELISA units (ELISA titer) represent the reciprocal of the serum dilution at which the absorbance at 414 nm is 1.0.   Data are redrawn from experiments shown by Richards et al. (1989).

relatively weak immunogens, and adjuvants have been widely proposed as adjuncts in the formulation of synthetic vaccines.   Liposomes are particularly useful as carriers of vaccines because additional adjuvants can be incorporated into the same liposomes that carry the antigen (reviewed by Alving, 1991).

Figure 1 demonstrates that a recombinant protein antigen (R32tet$_{32}$), carrying repeat sequence epitopes (asn-ala-asn-pro, or NANP; and asn-val-asp-pro, or NVDP) from the circumsporozoite protein of <u>Plasmodium</u> <u>falciparum</u>, was an extremely poor immunogen

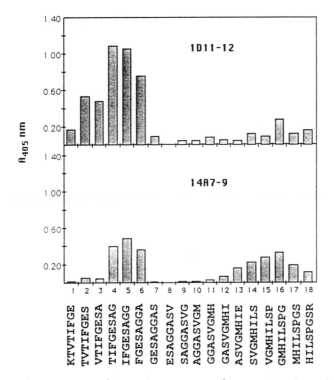

Figure 2. Epitope mapping of two murine monoclonal antibodies obtained after intraperitoneal immunization with liposomes containing an unconjugated synthetic peptide (25 amino acid residues) corresponding to the amino acid sequence surrounding the active site serine of *Torpedo californica* acetylcholinesterase. The liposomes also contained lipid A. Eighteen overlapping octapeptides were synthesized on polyethylene pins according to Geysen et al. (1984) and used in ELISA. The sequences of peptides are given at the bottom. From Ogert et al., 1990.

Phagocytosis of liposomes results in the accumulation and delivery of liposomal contents into intracellular vacuoles, including lysosomes. The liposomes and their contents are thought to be gradually degraded by lysosomal enzymes, and this has served as the basis for delivery of liposomal drugs for treatment of certain diseases in which lysosomes play a major role. Recently, we have proposed an intracytoplasmic pathway through which liposomal antigenic epitopes are processed by macrophages (Verma et al., 1991). Evidence for this proposed pathway is based on in monkeys, even when the antigen was adsorbed with aluminum hydroxide (alum) as an adjuvant. This formulation ("Falciparum Sporozoite Vaccine-1", or FSV-1) was also found to be poorly immunogenic in humans (Ballou et al., 1987). In contrast, inclusion of R32tet$_{32}$ in lipid A-containing alum-adsorbed liposomes, [L(Ag + Lipid A-44) + alum], resulted in a very potent immune response to the malarial antigen (Richards et al., 1989).

Unconjugated Peptides: Liposomes also may provide an interesting subtle advantage to vaccine formulation through an ability to induce conformational specificities of antibodies. An unconjugated 25 amino acid synthetic peptide containing epitopes surrounding the active site serine of *Torpedo californica* acetylcholinesterase was incorporated into liposomes containing lipid A and used as an immunogen to generate murine monoclonal antibodies. The specifities of the antibodies were examined by pin analysis according to the method of Geysen et al. (1984). As shown in Fig. 2, the antibodies reacted with the ends, but not the center, of the peptide. It was concluded that unique specificities were achieved with the liposomal antigen because of folding of the peptide within the lipid bilayer of the liposomes (Ogert et al., 1990).

INTRACELLULAR FATE OF LIPOSOMAL ANTIGEN IN MACROPHAGES

It is widely believed among immunologists that protein antigens must be processed and usually even partially degraded by antigen presenting cells in order to interact with the immune system in a form that is suitable for induction of an immune response. The major cells that are responsible for antigen presentation are macrophages, B lymphocytes, and dendritic cells. One of the earliest and most widely studied observations in the field of liposome-cell interactions is that liposomes are avidly ingested by macrophages. Because of this it has been proposed that macrophages serve as the primary antigen presenting cell for liposomal antigens (reviewed by Alving, 1991).

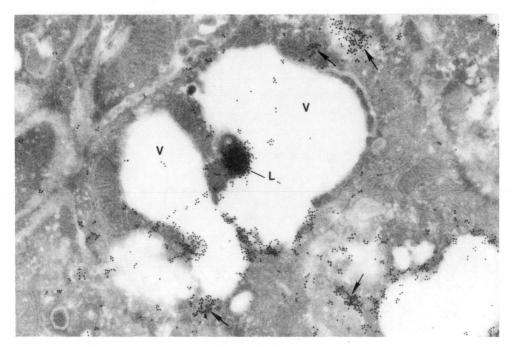

Figure 3. Immunogold electron microscopy of cultured bone marrow-derived macrophages 6 hours after phagocytosis of liposomes containing malaria antigen (R32NS1$_{81}$). The malaria antigen was detected by a specific monoclonal antibody to the antigen followed by treatment with gold-labelled second antibody. V, vacuole; L, liposome containing antigen. Four arrows indicate examples of locations of cytoplasmic antigen. See Verma et al. (1991) for further details.

electron microscopic observations such as that shown in Fig. 3.

Liposomal antigenic epitopes were identified with a monoclonal antibody, and the sites of the epitopes were determined with a gold-labelled second antibody (Fig. 3). After phagocytosis of liposomes a large number of gold-labelled spherules, presumably liposomes containing antigen, were observed in large, relatively clear, vacuoles in the macrophages. Liposomes were also found, but to a lesser extent under the conditions examined, in dense lysosomes.

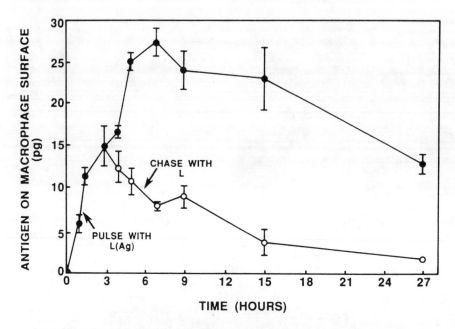

Figure 4. Kinetics of antigen expression on the macrophage surfaces as determined by a pulse-chase experiment. Macrophage cultures were incubated (pulsed) with liposomes containing antigen [L(Ag)] for 3 hours, followed by a 24 hour chase in which L(Ag) was replaced by liposomes lacking antigen [L]. Control cultures were incubated continuously for 27 hours with L(Ag). From Verma et al. (1991).

The observation that liposomes are accumulated within intracellular vacuoles is compatible with numerous similar previous observations. However, Fig. 3 also demonstrates that the liposomal antigenic epitopes were also detected in the cytoplasm of the cells. The epitopes were not present in the form of distinctive spherical accumulations that might suggest that they were still present within liposomes. Based on numerous electron microscopic fields that were similar to that shown in Fig. 3 we concluded that the liposomal antigenic epitopes had escaped from the vacuoles into the cytoplasm of the cells. The possibility was therefore raised that at least certain liposomal antigenic protein epitopes might be protected from complete degradation in lysosomal vacuoles.

What was the fate of liposomal antigenic protein epitopes that escaped from vacuoles into the cytoplasm? At the present time we do not have a definitive answer to that question, but we did make a further observation that might provide a clue. We discovered that shortly after phagocytosis of the liposomes containing antigen, the antigenic epitopes appeared on the surface of the macrophages. This was determined by ELISA using whole adherent macrophages as antigens. The ELISAs performed on whole cultured macrophages employed the same monoclonal antibody that was used for the immunogold electron microscopy. The antigenic epitopes were expressed on the surface of the macrophages within 15 min after phagocytosis of liposomes. As shown in Fig. 4, surface expression of antigenic epitopes reached a peak approximately 6 hours after initial incubation of the liposomes containing antigen with the macrophages.

In the experiment shown in Fig. 4, a pulse-chase experiment was performed in which liposomes lacking antigen ("empty" liposomes) were substituted for antigen-containing liposomes 3 hours after initiating the incubation with macrophages. The data indicate that antigen expression at the macrophage surface persisted, albeit at a declining level, over a period of 24 hours.

CONCLUSIONS: The data suggest that after phagocytosis of liposomes by macrophages, and accumulation of liposomes in vacuoles, there may be a directed flow of liposomal antigen out of the vacuoles into the cytoplasm. It does not appear that whole liposomes can escape in large quantities into the cytoplasm, but it is possible that the escaped antigen in the cytoplasm contains some associated liposomal lipid. We believe, although we do not yet have definitive proof, that the antigen that was originally present in the phagocytosed liposomes and that then escaped into the cytoplasm subsequently appeared at the surface of the macrophage. Regardless of the exact details involved in the translocation of liposomal antigen, this novel cytoplasmic pathway could represent a unique

202

mechanism by which phagocytic antigen presenting cells process liposomal antigens.

Acknowledgement: We thank SmithKline Beecham Pharmaceuticals, Swedeland, PA, for providing the R32tet$_{32}$ and R32NS1$_{81}$ antigens.

REFERENCES

Alving, C.R. (1991) J. Immunol. Meth. 140, 1-13
Ballou, W.R., Sherwood, J.A., Neva, F.A., Gordon, D.M., Wirtz, R.A., Wasserman, G.F., Diggs, C.L., Hoffman, S.L., Hollingdale, M., Hockmeyer, W.T., Schneider, I., Young, J.F., Reeve, P., and Chulay, J.D. (1987) Lancet i, 1277-1280
Geysen, H.M., Meloen, R.H., and Barteling, S.J. (1984) Proc. Natl. Acad. Sci. U.S.A. 81, 3998-4002
Ogert, R.A., Gentry, M.K., Richardson, E.C., Deal, C.D., Abramson, S.N., Alving, C.R., Taylor, P., and Doctor, B.P. (1990) J. Neurochem. 55, 756-763
Richards, R.L., Swartz, G.M., Jr., Schultz, C., Hayre, M.D., Ward, G.S., Ballou, W.R., Chulay, J.D., Hockmeyer, W.T., Berman, S.L., and Alving, C.R. (1989) Vaccine 7, 506-512
Verma, J.N., Wassef, N.M., Wirtz, R.A., Atkinson, C.T., Aikawa, M., Loomis, L.D., and Alving, C.R. (1991) Biochim. Biophys. Acta 1066, 229-238

Progress in Membrane Biotechnology
Gomez-Fernandez/Chapman/Packer (eds.)

TREATMENT OF TUMORS USING CYTOSTATIC-CONTAINING LIPOSOMES

G. Storm, U.K. Nässander, P.A. Steerenberg[1] and D.J.A. Crommelin

Department of Pharmaceutics, University of Utrecht, P.O. Box 80.082, 3508 TB Utrecht, The Netherlands, [1]Department of Pathology, National Institute of Public Health and Environmental Protection, P.O. Box 1, 3720 BA Bilthoven, The Netherlands

SUMMARY: Research on liposome-encapsulated antitumor drugs has proceeded at a steady pace during the last decade. Most progress in this field has been in connection with the use of anthracyclines and platinum compounds. It is imperative that an intended therapeutic application of liposomes should be well matched with the liposome behavior in vivo. Aspects of liposome biodistribution relevant to cancer chemotherapy are briefly addressed, followed by a discussion of areas in chemotherapy with a strong rationale for the use of liposomes.

INTRODUCTION

Liposomes consist of one or more concentric phospholipid bilayers enclosing an equal number of aqueous compartments in which water-soluble substances can be entrapped, whereas lipid-soluble substances can be incorporated into the bilayers. Various techniques have been described to prepare liposomes of different properties, such as size (25 nm to 20 µm), number of lamellae, charge and fluidity (Gregoriadis, 1984; Lichtenberg et al., 1988; Nässander et al., 1990). Liposomes have been widely used as model membranes and their physicochemical properties have therefore been studied extensively. More recently, they have become important tools for the study of membrane-mediated processes (e.g., membrane fusion), catalysis of reactions occurring at interfaces, and energy conversion. Besides, liposomes are currently under investigation as carrier system for drugs. Their application as drug carriers in cancer chemotherapy is subject of this contribution.

The narrow therapeutic window of most antitumor drugs has prompted a search, not only for new drugs, but also for innovative ways to improve therapeutic utilization of existing drugs. Prominent among these approaches has been the employment of drug delivery technology such as liposomes. Research on liposome-incorporated antitumor drugs has proceeded at a steady pace over the last few years. While a variety of agents have been incorporated, most progress has been in connection with the use of anthracyclines and platinum compounds. Clinical investigations are currently underway and some reasonably promising results have been attained (Daoud et al., 1989).

We will first briefly address some aspects of liposome behavior in vivo relevant to cancer treatment. This will be followed by a section which attempts to identify those areas in chemotherapy with a strong rationale for the use of liposomes. Finally, we will discuss the potential usfulness of liposomes for intraperitoneal cancer therapy as this example can be helpful in gaining a perspective of the therapeutic possibilities of liposomal delivery of antitumor drugs.

IMPLICATIONS OF BIOLOGICAL FACTORS FOR DRUG DELIVERY TO TUMOR CELLS

In the initial literature on liposome-based delivery of antitumor drugs, enthusiasm was generated primarily by the idea that liposomes might be used for specific delivery of the antitumor drug to tumor cells. However, this was followed by a period of pessimism as researchers obtained a better understanding of the biological barriers counteracting an adequate in vivo performance of liposomal antitumor drugs (for reviews of liposome behavior in vivo see Poznanski and Juliano, 1984, Senior, 1987, Juliano, 1988, Crommelin and Storm, 1990, and Nässander et al., 1990).

In particular it became evident that liposomes were rapidly cleared from the bloodstream and accumulated in the phagocytic cells of the reticuloendothelial system (RES), also referred to as mononuclear phagocyte system (MPS), especially liver Kuppfer cells and splenic macrophages. This separates the liposomal drug from the tumor and may induce adverse effects towards the RES, a vital component of host defense. Abundance of phagocytic macrophages and a rich blood supply are the primary reasons for the predominant uptake of liposomes by liver and spleen.

Considering their relatively large size, liposomes have only a very limited ability to reach extravascular tissues. Only small-sized liposomes (less than 100 nm) might extravasate through endothelial gaps (up to 100 nm in size) in sinusoidal capillaries (Poste et al., 1982; Scherphof et al., 1983). Thus, it also became clear that liposomes, or at least those with a larger size, do not penetrate the barrier represented by the capillary endothelium within the blood supply of solid tumors, again preventing direct access of liposomal drug to the tumor cells.

Various factors can impair the integrity of the liposomes in the circulation resulting in leakage and premature loss of drug to be delivered. For example, the activation of complement can lead to their destruction. Moreover, serum components like high-density lipoproteins (HDL), serum albumin, the alpa- and beta-globulins and naturally occurring antiphospholipid antibodies can interact with liposomes and may affect their stability considerably. Exchange can occur between the cholesterol of liposomes and that of blood cells and lipoproteins. It is possible to promote the serum stability of liposomes by an appropriate choice of lipid constituents, e.g. by inclusion of cholesterol, use of sphingmyelin and use of phospholipids with long, fully saturated acyl chains. Although serum stability is an important parameter, this does not necessarily mean that stable liposomes are more effective carriers in general. For some valuable applications slow release from the liposomes circulating in the bloodstream or present at the target site is desired. In such cases, liposome stability should not inhibit extracellular release of the drug (see below).

In more recent years, the overly pessimistic view has been superseded by a more balanced approach. A major consequence of drug encapsulation in liposomes is a changed pharmacokinetic and tissue distribution profile of the drug. The changes in pharmacokinetics are generally characterized by an increased area under the curve with a decreased renal clearance, as compared to the free drug. Tissue distribution studies have shown that drug levels in liver and spleen are sustantially raised, while lower levels are generally observed in thoses tissues separated from the intravascular compartment by a continuous capillary endothelium and an intact basement membrane. The question is whether we can take advantage of these pharmacological changes obtained with liposomes in order to achieve a higher therapeutic index.

RATIONALES FOR THE ENCAPSULATION OF CYTOSTATICS IN LIPOSOMES

When drug carriers such as liposomes are considered for more effective drug delivery in cancer chemotherapy, two basic issues are involved: targeting and sustained release of drugs. Targeting encompasses two different forms. One is *passive targeting* which relates to the natural distribution patterns of carrier systems. A more specific form of drug targeting is *active targeting*, involving the linkage of some type of a ligand (e.g., a monoclonal antibody) to the drug carrier. Targeting is attained by the recognition at the molecular level through the direct and specific interaction between the recognition site of the drug carrier and the receptor on the target cell. *Sustained release* refers to the potential of liposomes to provide a 'depot' for release of an encapsulated drug over an extended period of time.

These issues, sustained release and targeting, provide the rationales for the use of liposomes for the enhancement of the therapeutic index of antitumor agents. Recent studies show that liposomes can reduce certain types of drug-related toxicity with preservation of antitumor activity. This is well founded by preclinical experimentation with regard to doxorubicin (DXR), daunorubicin and actinomycin (Mayhew and Papahadjopoulos, 1983; Gabizon, 1989; Nässander et al., 1990). At least two factors may contribute to this increase in therapeutic index: site avoidance of drug sensitive tissues, such as the heart muscle in the case of anthracyclines, and sustained release of drug avoiding the high plasma concentrations of bolus injection of free drug. It has been shown for DXR that the encapsulated drug can be released directly from liposomes being present in the bloodstream but also indirectly from the RES after uptake and processing by macrophages (Storm et al., 1987, 1988, 1989a,b,c). Peak concentrations of 'free' DXR in organs which are particularly sensitive to the toxic action of the drug, like the heart, are avoided, while apparently the prolonged presence of 'free' (i.e. non-liposomal) drug levels in the blood can result in sufficient exposure levels for tumor cells.

The liver is a common site of metastatic spread with a most detrimental effect on the prognosis of cancer patients. Passive targeting of liposome-encapsulated drugs to the liver represents a distinct and attractive possibility to increase drug concentration in tumor cell colonies residing in this organ. This approach is supported by preclinical evidence in various mouse metastatic tumor models using DXR-liposomes (Gabizon et al., 1983;

Mayhew et al., 1983; Gabizon et al., 1985). Liposomal DXR was found to be up to 100-fold more active than free DXR in two murine models of liver metastases. Penetration of small-sized liposomes through sinusoidal capillaries (see previous section) and RES-mediated sustained release effects might have been involved in producing the positive outcome.

As discussed in the previous section, systemic delivery of liposomes to increase drug delivery to tumor cells faces two serious obstacles: fast and predominant uptake by the RES, and little or no ability of particulate carriers to extravasate other than through sinusoidal capillaries. Several groups explored the idea of avoiding uptake by the RES (Davis and Illum, 1986; Allen and Chonn, 1987; Gabizon and Papahadjopoulos, 1988). Liposomes which are capable of evading the RES would provide two important advantages. First, the liposome circulation time in the blood is increased. This would prolong the time frame available for slow drug release in the bloodstream and would also improve the prospect of directing liposomes to tissues other than those rich in RES cells. The other advantage may be decreased loading of the RES, thus avoiding blockade and impairment of an important host defense system (Storm et al., 1991). One of the early approaches which have been followed to increase liposome circulation time was "blocking" of the phagocytic uptake mechanisms of the RES by predosing with high doses of liposomes or other particles. Another approach was based on improving liposome stability in plasma (Senior, 1987). Although both approaches resulted in reduced clearance rates of circulating liposomes, they were not very successful in achieving substantial liposome uptake by tissues outside the RES (Poste, 1983). Profitt et al. (1983) reported a 50% increase in the uptake of radiolabeled liposomes in tumors of mice with a suppressed RES system by predosing with "cold" liposomes compared to "non-suppressed" animals. However, RES suppression seems to be of limited value in clinical practice since the RES plays an important role in the body's defense system. Thus, it is not feasible to close it down for prolonged periods of time.

However, recent evidence (Gabizon and Papahadjopoulos, 1988; Allen et al., 1989; Woodle et al., 1991; Bakker-Woudenberg et al., 1991) suggests that longer circulation circulation of liposomes still may offer viable therapeutic opportunities. It was found that inclusion of monosialoganglioside, phosphatidylinositol, sulfatides, or polyethyleneglycol prolongs the circulation times of liposomes, particularly in combination with liposomal formulations with rigid bilayers(e.g., containing sphingomyelin, distearoylphosphatidylcholine, cholesterol, or other rigidifying phospholipids). Allen (1989) showed that as much as 50% of the injected dose remains in the circulation for 8 h with up to 20% of the injected dose still circulating after 24 h. It has been proposed that the phenomenon of prolonged circulation and concomitant reduced RES uptake might be related to a decreased opsonization, probably obtained by mimicking important properties of the outer monolayer of red blood cells (Allen and Chonn, 1987). Clearly, such long circulating liposomes have a far greater opportunity than conventional liposomes to be distributed to non-RES sites. A remarkably increase in tumor uptake (up to 25-fold as compared to conventional liposomes) was observed when such RES-avoiding liposome formulations with a mean size around 100 nm were tested in mice bearing an implanted intramuscular tumor (Gabizon and Papahadjopoulos, 1988; Gabizon et al., 1990). Concomitantly, there was a decrease by a factor 4 of the recovered dose localizing in the liver and spleen, the major RES organs. The mechanism of liposome accumulation in areas of implanted tumors and their detailed localization remains unclear at present. The permeability of tumor vasculature is generally increased as compared to normal tissues (Jain and Gerlowski, 1986), although this varies among tumors. The broad heterogeneity of tumors with regard to microvascular architecture (Shubik, 1982) makes it difficult to make any predictions on accessibility and uptake of liposomes. In this context, small-sized liposomes would be the more advantageous ones for drug delivery to tumors.

From above it is clear that RES capture as well as the barrier function of the endothelium are major factors counteracting the extravascular disposition of liposomes after intravenous administration. Therefore, the intracavitary administration of liposomal anticancer agents as therapy for tumors confined to body cavities has significant theoretical appeal. Pharmacokinetic evaluation of several commonly used antineoplastic drugs has confirmed modeling predictions that suggested a significantly greater exposure of the cavity to active drug than the plasma when the drugs are delivered directly into the body cavity (Markman, 1986a). Liposome encapsulation may provide additional advantages such as increased area under the curve at the site of action and reduced systemic exposure of the drug. In the next section, we will deal in more detail with the use of liposomes for intracavitary therapy of malignancies, such as ovarian carcinoma, which reside principally in the peritoneal cavity.

INTRAPERITONEAL ADMINISTRATION OF LIPOSOME-ENCAPSULATED CYTOSTATICS

Intraperitoneal (i.p.) administration of chemotherapeutic agents has been used for many years as a way of increasing the delivery of drugs to tumors (e.g., ovarian carcinoma) located in the peritoneal cavity (Markman et al., 1985, Markman, 1986b; Markman et al., 1986; Howell and Zimm, 1988; Los, 1990). Cis-diamminedichloroplatinum (cDDP), cytosine arabinoside (Ara-C), and bleomycin are examples of i.p. administered drugs which were already successfully applied in clinical settings. In comparison with the i.v. route of administration the potential advantages of i.p. therapy are the avoidance of high toxic drug plasma levels and an increased (local) exposure of tumors to anticancer drugs. Whether this increased exposure will result in an improved therapeutic effect depends on free-surface diffusion into the tumor. In addition, after i.p. administration, a substantial fraction of the drug will finally reach the blood, which enables the drug to affect the tumor via the capillary route.

Several studies have been performed in order to investigate the fate of liposomes after i.p. administration (Parker et al., 1981; Senior and Gregoriadis, 1982; Ellens et al., 1983; Poste, 1983; Hirano and Hunt, 1985; Nässander, 1991). Like other high molecular weight and particulate materials, liposomes are removed from the peritoneal cavity via the lymphatics of the diaphragm. This may be therapeutically beneficial as tumor cells often metastasize through lymphatic channels. Taking into account the diameter of the lymphatic capillaries lining the diaphragm, particles larger than about 23 µm are not expected to enter the lymphatics (Allen, 1956). After i.p. administration of liposomes with a mean particle size of 19 ± 6 µm into mice, it has been shown that liposomal entrapment of Ara-C prolongs the half-life of the drug in the peritoneal cavity by 79-fold (Kim and Howell, 1987; Kim et al., 1987). The i.p. administered liposomes remained in the peritoneal cavity with a half life of 21 h. The prolonged retention time of Ara-C in the peritoneal cavity resulted in better therapeutic effects on i.p. inoculated L1210 cells, as compared to free drug. Sur et al. (1983) reported on a liposome-mediated increase of antitumor activity and reduction of toxicity of i.p. administered cDDP in mice bearing Ehrlich ascites carcinoma. The ability of i.p. administered liposomes to prolong the retention of drug has also been demonstrated by our research group (Nässander, 1991). However, antitumor effects of liposome-encapsulated cytostatics (cDDP and DXR) were inferior to those of the free drugs in two intraperitoneal tumor systems. Experimental evidence was collected which pointed to insufficient drug release as a likely explanation for the disappointing results. Data obtained on the release kinetics of drug and other entrapped substances from the liposomes in serum and ascites strongly indicated that manipulation of the liposomal lipid composition can be an important tool for improving the antitumor activity of i.p. injected liposomal drugs. For a full evaluation of the potential of cyostatic laden liposomes, however, not only the antitumor activity, but also the toxic effects should be taken into account. Considering that liposomes may also provide a considerable reduction in dose-limiting toxicity, the final conclusion on the therapeutic advantage of this approach should be drawn on the basis of antitumor activity results obtained with equitoxic doses (maximum tolerated doses).

In order to obtain additional therapeutic benefits, site-specific drug delivery into this anatomically isolated cavity by using immunoliposomes (liposomes to which tumor-specific antibodies are covalently linked) might be favorable. Hashimoto et al. (1983) injected i.p. mammary tumor cells in mice. Twenty-four hours later actinomycin D-containing immunoliposomes were injected i.p.; free drug and "naked" liposomes mixed with the free antitumor drug were used as controls. At the dose level used 100% survival was observed after injection of drug-containing immunoliposomes; all control animals died rapidly. Another interesting study has been performed by Wang and Huang (1987) who demonstrated target cell-specific delivery of a liposome-encapsulated plasmid to i.p. located RDM4 lymphoma cells after i.p. administration.

Our group is currently evaluating the usefulness of cytostatic-containing immunoliposomes in the therapy of human ovarian cancer (Nässander, 1991). The successful treatment of this tumor, being the leading cause of death in women with a gynecologic malignancy, is an ongoing oncological challenge. The main reason for the high mortality is the late occurrence of symptoms, resulting in an advanced diseased state at the time of diagnosis. Even with aggressive combined modality treatment, the prognosis for these patients is poor. Since ovarian cancer remains confined to the peritoneal cavity virtually throughout its entire clinical course, this type of cancer is an attractive candidate for i.p. chemotherapy. Local instillation of anticancer agents has been used with some success, although systemic absorption of drug, as well as local drug effects, can produce substantial toxicity. Encapsulation of cytostatics in liposomes might help to solve these problems (Straubinger et al., 1988). Forssen and Tökes (1983) reported that the local inflammation produced by DXR is reduced by encapsulation in liposomes. Our recent results show that immunoliposome-encapsulated drug resides much longer in the peritoneal cavity of athymic nude mice bearing an i.p. growing human ovarian carcinoma than free drug. The antibody-mediated binding of immunoliposomes to the tumor cells was rapid: 30 minutes after i.p. injection approximately 70% of the injected dose of immunoliposomes was associated with the tumor cells while for nontargeted liposomes (bearing no antibody) only a value of approximately 3% was obtained. At 2 h after i.p. injection, a maximal binding level of 84% was achieved in case of the specific immunoliposomes whereas the binding level of nontargeted liposomes was still about 3%. Twenty-four hours after injection still about 83% of the injected dose of specific immunoliposomes was associated with the tumor cells compared to about 10% of the injected dose of nontargeted liposomes. Accordingly, nontargeted liposomes disappeared from the peritoneal cavity much faster that the specific immunoliposomes. At present, experiments are being conducted to demonstrate and optimize therapeutic efficacy in vivo. Whether the encapsulated cytostatic drug will exert its therapeutic effect depends on the endocytotic capacity of the tumor cells and/or the integrity of the immunoliposomes used. Temporarily high concentrations of drug released from bound immunoliposomes in close proximity of the tumor cells might be very effective.

The question arises how far is the liposome-based approach of ovarian carcinoma chemotherapy away from its introduction into the clinic. Recently, a preliminary Phase I and II clinical trial was carried out by Delgado et al. (1989) to study the i.p. administration of liposome-encapsulated DXR in 15 patients with refractory advanced ovarian cancer. DXR was formulated in liposomes consisting of cardiolipin, phosphatidylcholine, cholesterol, and stearylamine. No antibodies were coupled to the liposomes. Encouraging results were obtained. Liposomal DXR was better tolerated than free drug. Patients treated with liposomal DXR received more than double the dosage of DXR which, when administered in free form, resulted in unacceptible local toxicity (Ozols et al., 1982). Three patients with a small tumor burden at the initiation of therapy showed a partial response. Five of the six nonresponders had bulky disease at the beginning of the study. In their progress report, the maximum tolerated dose was not reached yet. The

investigators are currently awaiting the response data from patients with a
small volume of disease at the higher drug doses.

REFERENCES

Allen, L. (1956) Anat. Rec. 124, 639-657.

Allen, T.M., and Chonn, A. (1987) FEBS Lett. 223, 42-46.

Allen, T.M. (1989) In: Liposomes in the Therapy of Infectious Diseases and
    Cancer (I.J.Fidler and G. Lopez-Berestein, Eds.), Alan R. Liss, New York,
    pp. 405-415.

Allen, T.M., Hansen, C., and Rutledge, J. (1989) Biochim. Biophys. Acta 981,
    27-35.

Bakker-Woudenberg, I.A.J.M., Lokerse, A.F., Ten Kate, M.T., and Storm, G.
    Submitted for publication.

Crommelin, D.J.A., and Storm, G. (1990) In: Comprehensive Medicinal Chemistry
    (C. Hansch, P.G. Sammes and J.B. Taylor, Eds.), Vol. 5, Pergamon Press,
    Oxford, pp. 57-77.

Daoud, S.S., Hume, L.R., and Juliano, R.L. (1989) Adv. Drug Delivery Rev. 3,
    405-418.

Davis, S.S., and Illum, L.L. (1986) In: Site-specific Drug Delivery (E.
    Tomlinson and S.S. Davis, Eds.) John Wiley & Sons, Chichester, pp. 93-110.

Delgado, G., Potkul, R.K., Treat, J.A., Lewandowski, G.S., Barter, J.F.,
    Forst, D., and Rahman, A. (1989) Am. J. Obstet. Gynecol. 160, 812-819.

Ellens, H., Morselt, H.W.M., Dontje, B.H.J., Kalicharan, D., Hulstaert, C.E.,
    and Scherphof, G.L. (1983) Cancer Res. 43, 2927-2934.

Forssen, E.A., and Tökes, Z.A. (1983) Cancer Treat. Rep. 67, 481-484.

Gabizon, A., Goren, D., Fuks, Z., Dagan, A., Barenholz, Y., and Meshorer, A.
    (1983) Cancer Res. 43, 4730-4735.

Gabizon, A., Goren, D., Fuks, Z., Meshorer, A., and Barenholz, Y. (1985) Br.
    J. Cancer 51, 681-689.

Gabizon, A., and Papahadjopoulos, D. (1988) Proc. Natl. Acad. Sci. USA 85,
    6949-6953.

Gabizon, A. (1989) In: Drug Carrier Systems (F.H. Roerdink and A.M. Kroon,
    Eds.), John Wiley & Sons Ltd., pp. 185-211.

Gabizon, A., Price, D.C., Huberty, J., Bresalier, R.S., and Papahadjopoulos,
    D. (1990) Cancer Res. 50, 6371-6378.

Gregoriadis, G. (Ed.) (1984) Liposome Technology, Vols. I, II and III, CRC
    Press, Inc., Boca Raton, Florida.

Hashimoto, Y., Sugawara, M., Masuko, T., and Hojo, H. Cancer Res. 43, 5328-
    5334.

Hirano, K., and Hunt, C.A. (1985) J. Pharm. Sci. 74, 915-921.

Howell, S.B., and Zimm, S. (1988) Acta Chir. Scand. Suppl. 541, 16-21.

Jain, R.K., and Gerlowski, L.E. (1986) Crit. Rev. Oncol. Hematol. 5, 115-170.

Juliano, R.L. (1988) Adv. Drug. Deliv. Rev. 2, 31-54.

Kim, S., and Howell, S.B. (1987) Cancer Treatm Rep. 71, 705-711.

Kim, S., Kim, D.J., and Howell, S.B. (1987) Cancer Chemother, Pharmacol. 19,
    308-310.

Lichtenberg, D., and Barenholz, Y. (1988) In: Methods of Biological Analysis
    33 (D. Glick, Ed.), John Wiley & Sons, New York, pp. 337-461.

Los, G. (1990) In: Experimental Basis of Intraperitoneal Chemotherapy (Ph. D.
    Thesis), Centrale Huisdrukkerij Vrije Unversiteit, Amsterdam, The
    Netherlands.

Markman, M., Cleary, S., Lucas, W., and Howell, S.B. (1985) J. Clin. Oncol.
    3, 925-931.

Markman, M. (1986a) Am J. Med. Sci. 291, 175-179.

Markman, M. (1986b) Cancer Treatm. Rev. 13, 219-242.

Markman, M., Cleary, S., Lucas, W., Weiss, R., and Howel, S.B. (1986) Cancer Treatm. Rep. 70, 755-760.

Mayhew, E., Rustum, Y., and Vail, W.J. (1983) Cancer Drug Deliv. 1, 43-58.

Mayhew, E., and Papahadjopoulos, D. In: Liposomes (M.J. Ostro, Ed.), Marcel Dekker, New York, pp. 289-341.

Nässander, U.K., Storm, G., Peeters, P.A.M., and Crommelin, D.J.A. (1990) In: Biodegradable Polymers as Drug Delivery Systems (M. Chasin and R. Langer, Eds.), Marcel Dekker, New York, pp. 261-338.

Nässander, U.K. (1991) In: Liposomes, Immunoliposomes, and Ovarian Carcinoma (Ph. D. Thesis), Krips Repro, Meppel, The Netherlands.

Ozols, R.F., Young, R.C., Speyer, J.L., Sugarbaker, P.H., Greene, R., Jenkins, J., and Myers, C.E. (1982) Cancer Res. 42, 4265-4269.

Parker, R.J., Hartman, K.D., and Sieber, S.M. (1981) Cancer Res. 41, 1311-1317.

Poste, G., Bucane, C., Raz, A., Bugelski, P., Kirsh, R., and Fidler, I.J. (1982) Cancer Res. 42, 1412-1422.

Poste, G. (1983) Biol. Cell. 47, 19-38.

Poznanski, M.J., and Juliano, R.L. (1984) Pharmacol. Rev. 36, 277-336.

Profitt, R.T., Williams, L.E., Presant, C.A., Tin, G.W., Uliana, J.A., Gamble, R.C., and Baldeschwieler, J.D. (1983) Science 220, 502-505.

Scherphof, G., Roerdink, F., Dijkstra, J., Ellens, H., De Zanger, R., and Wisse, E. (1983) Biol. Cell 47, 47-58.

Senior, J., and Gregoriadis, G. (1982) Life Sci. 30, 2123-2136.

Senior, J. (1987) CRC Crit. Rev. Therapeut. Drug Carrier Syst. 3, 123-193.

Shubik, P. (1982) J. Cancer Res, Clin. Oncol. 103, 211-226.

Storm, G., Steerenberg, P.A., Roerdink, F.H., De Jong, W.H., and Crommelin, D.J.A. (1987) Cancer Res. 47, 3366-3372.

Storm, G., Steerenberg, P.A., Emmen, F., Van Borssum-Waalkes, M., and Crommelin, D.J.A. (1988) Biochim. Biophys. Acta 965, 136-145.

Storm, G., Regts, J., Beijnen, J.F., and Roerdink, F.H. (1989a) J. Lip. Res. 1, 195-210.

Storm, G., Nässander, U.K., Steerenberg, P.A. Roerdink, F.H. De Jong, W.H. and Crommelin, D.J.A. (1989b) In: Liposomes in the Therapy of Infectious Diseases and Cancer (I.J.Fidler and G. Lopez-Berestein, Eds.), Alan R. Liss, New York, pp. 105-116

Storm, G., Van Hoesel, Q.G.C.M., De Groot, G., Kop, W., Steerenberg, P.A., and Hillen, F.C. (1989c) Cancer Chemother. Pharmacol. 24, 341-348.

Storm, G., Oussoren, C., and Peeters, P.A.M. (1991) In: Membrane Lipid Oxidation, Vol. III (C. Vigo-Pelfrey, Ed.), CRC Press Inc., Boca Raton, Florida, pp. 239-263.

Straubinger, R.M., Lopez, N.G., Debs, R.J., Hong, K., and Papahadjopoulos, D. Cancer Res. 48, 5237-5245.

Sur, B., Ray, R.R., Sur, P., and Roy, D.K. (1983) Oncology 40, 372-376.

Wang, C., and Huang, L. (1987) Proc. Natl. Acad. Sci. USA 84, 7851-7855.

Woodle, M.C., Storm, G., Newman, M.S., Jekot, J., Collins, L.R., Martin, F.J., and Szoka, F.C. Submitted for publication.

Progress in Membrane Biotechnology
Gomez-Fernandez/Chapman/Packer (eds.)
© 1991 Birkhäuser Verlag Basel/Switzerland

LIPOSOME ENCAPSULATED HEMOGLOBIN; IN-VIVO EFFICACY OF A SYNTHETIC RED CELL SUBSTITUTE.

Alan S. Rudolph[1], Beth Goins[2], Frances Ligler[1], Richard O. Cliff[2], Helmut Spielberg[3], Peter Hoffman[3], William Phillips[4], Robert Klipper[4]

[1]Center for Biomolecular Science and Engineering, Code 6090, Naval Research Laboratory, Washington, DC, [2]Geo-Centers Inc., Fort Washington, MD, [3]Department of Biochemistry, Georgetown University, Washington, DC, [4]Department of Nuclear Medicine, University of Texas Health Science Center at San Antonio, San Antonio, TX

Summary: The encapsulation of hemoglobin within liposomes is one strategy toward the development of an oxygen-carrying resuscitative fluid. The current status of the liposome encapsulated hemoglobin (LEH) project has evolved from considerations of production methods and in-vivo efficacy. Currently, LEH can be manufactured in large scale (liter quantities) using high pressure sheer and is comprised of 50 mole percent distearoyl phosphatidylcholine, 38% cholesterol, 10% dimyristoyl phosphatidylglycerol, and 2% alpha-tocopherol with the encapsulation of bovine hemoglobin. Previous efforts to ascertain LEH efficacy have focused on circulation persistence in total isovolemic exchange transfusion models. More recently we have concentrated on introducing clinically relevant doses of LEH to investigate the hemodynamic, hematologic, immunologic, and pathologic consequences of LEH normovolemic infusion. Studies with radiolabeled LEH also reveal circulation persistence and biodistribution of LEH in the rabbit with the majority of LEH accumulation over 24 hours in the liver and spleen, primary organs of the reticuloendothelial system (RES). Pathology studies of mice given LEH and sacrificed up to 4 weeks after infusion show that LEH is handled by the RES and cleared, with normal organ histology observed at 1-2 weeks. There is no accumulation of LEH in the kidney, a potential site of hemoglobin-induced toxicity. The long-term storage of LEH has been accomplished by using water replacement molecules to preserve LEH in the freeze-dried state. Comparison of refrigerated fresh LEH and lyophilized and reconstituted LEH preparations in these same animal models shows the utility of this storage technology.

INTRODUCTION:

The development of a viable blood substitute is currently driven by the often insufficient reserves of banked blood and the need for an oxygen carrying resuscitative fluid that is free of potential transmittable viral antigens. Hemoglobin-based substitutes that are dependent on isolation of hemoglobin from outdated blood supplies also suffer from unpredictable supply. The Navy has been committed to the development of an encapsulated hemoglobin-based artificial oxygen-carrying fluid, liposome encapsulated hemoglobin (LEH). The philosophy of using liposomes as a vehicle for the delivery of hemoglobin is based on the pharmacokinetics of liposomes (and the possibility of long circulation half-lives), the known biocompatability and metabolism of liposomes which contain natural phospholipids and sterols, and the potential for reducing hemoglobin induced toxicity by presentation in an encapsulated form. The development of LEH over the years has focused on process engineering issues (Beissinger, et al., 1986, Jopski, et al., 1989), in-vitro oxygen-carrying capacity and kinetics (Miller, 1981, Hunt & Burnette, 1983, Gaber & Farmer, 1984, Farmer, et al., 1987, Farmer, et al., 1989), efficacy in exchange transfusions (Djordjevich, et al., 1982, Djordjevich, et al., 1983, Hunt & Burnette, 1983, Hunt, et al., 1985, Djordjevich, et al., 1987, Farmer, et al., 1989, Ligler, et al., 1989), and long-term storage by freeze-drying (Rudolph, 1988, Rudolph & Cliff, 1990).

The first obstacle toward the development of LEH was finding a formulation that was both compatible with significant circulation persistence as well as amenable to large scale production methods. LEH formulation currently consists of 50 mole% distearoyl phosphatidylcholine (DSPC), 38 mole% cholesterol, 10 mole% dimyristoyl phosphatidylglycerol (DMPG), and 2 mole% alpha-tocopherol. In-vivo half-life measurements in small rodents generated by following the disappearance of a LEH pellet (visually observed directly above the red cell fraction) indicate that this formulation has a dose dependent persistence which is longer (5-8 hours) than an identical dose of LEH prepared with the same mole%

of dimyristoyl phosphatidylcholine instead of the distearoyl lecithin (Gaber & Farmer, 1984, Farmer, et al., 1989). The circulation persistence of LEH with distearoyl phosphatidylcholine as the major phospholipid at 25% normovolemic infusion was 15-20 hours measured in small rodents (Gaber & Farmer, 1984, Farmer, et al., 1989, Ligler, et al., 1989). This formulation of LEH also withstood large scale preparative techniques such as the use of high pressure hydrodynamic sheer devices which are used for the large scale preparation of LEH (liter quantities can be made in a day).

The in-vitro parameters indicate that encapsulation of Hb into liposomes does not alter its ability to perform its cooperative loading and unloading of oxygen or carbon dioxide, with the binding and release kinetics determined by LEH size (surface to volume ratio) (Farmer, et al., 1989). Coencapsulants which can modify the oxygen-carrying capacity of the hemoglobin such as the organic phosphates, diphosphoglycerate (DPG) and pyridoxal-5-phosphate (P5P) have been shown to modify the oxygen-carrying capacity of LEH with the long-term success of these agents determined by the permeability coefficient across the liposome bilayer (Farmer, et al., 1989).

In-vivo studies of LEH began as large scale preparations became available. The initial studies were designed to demonstrate that LEH could support life in a scenario where death was eminent. This was accomplished by total exchange transfusion of LEH solutions in anesthetized rats. Male rats were exchanged isovolemically down to red cell hematocrit levels of 3% at which time all of the animals expired (n=6). Those animals isovolemically exchanged with LEH to 40% of the original circulating volume showed normal blood pressure, heart rate, systemic O_2, and blood gases following the infusion (Ligler, et al., 1989, Farmer, et al., 1989). The animals recovered from the anesthetic, and survived between 12-24 hours post-exchange (n=4). The demonstration of efficacy in this scenario was determined at the point at which the circulating level of red cells was not sufficient to support life and the presence of LEH resulted in normal oxygen levels, and the sustenance of life.

Exchange studies with LEH at doses which are clinically relevant, as well as more in depth studies of the *in-vivo* consequences of LEH administration have been the focus of our more recent research efforts (Rabinovici, et al., 1989, Rabinovici, et al., 1990a, Rabinovici, et al., 1990b). The purpose of the present report is to review recent efficacy data in small animal models and to present preliminary data on the immunologic, pathologic and hematologic consequences of LEH infusion in mice.

MATERIALS AND METHODS:

LEH manufacture and quality control: LEH is prepared by first solubilizing the liposome components in chloroform. Solvent evaporation results in the formation of a film to which a buffer solution is added. The resultant suspension is then processed through a high pressure (30-60 psi), hydrodynamic sheer apparatus (Microfluidics Corp., Boston, MA). The preparation is then lyophilized which results in the formation of a white powder. This step was added as hydration with hemoglobin in more efficient if the powder is hydrated as opposed to an organic film. A concentrated hemoglobin solution (20-24 gram percent) is added to the white powder and hydration is allowed to proceed overnight at $4^{\circ}C$. The multilamellar dispersion is then processed again through the hydrodynamic sheer apparatus to create LEH in the desired size range. LEH can be concentrated by either centrifugation or ultrafiltratiuon and teh unencapsulated hemoglobin recycled. The size is monitored using quasi-elastic light scattering methods. The exact size distribution depends on the pressure used during the process and the ratio of hemoglobin to lipid used in the preparation. The size distribution of LEH usually consists of a bimodal distribution with 60-80% of the population with a mean diameter of 0.2-0.4 microns, and the remaining fraction 1-2 microns.

In-vitro analysis of LEH consists of determining the hemoglobin concentration and oxidation state (methemoglobin), oxygen carrying capacity, endotoxin level, and sterility.

Intravesicular hemoglobin concentration and methemoglobin levels are determined using a colorimetric assay (Tomita, et al, 1968, Benesh & Yung, 1973). The hemoglobin concentration in LEH is ultimately determined by the final concentration of LEH in solution. At a lipid concentration of 17 mg/ml LEH lipid, there is 10-15 gram percent hemoglobin depending on the encapsulation efficiency. The methemoglobin levels of LEH depend on the starting concentration of methemoglobin from the manufacturer (Biopure Corp., Boston, MA). We have reported the use of antioxidants to reduce the level of methemoglobin before LEH production (Stratton, et al., 1988). After LEH manufacture, methemoglobin levels increase 3-8%. Oxygen-carrying capacity of LEH (P_{50}) is determined using a Hemox Analyzer (TCS Medical Products, Huntingdon Valley, PA). The P_{50} of LEH is usually 18-22 in phosphate buffered saline but can be increased by coencapsulation of organic phosphates, diphosphoglyerate or pyridoxal-5-phosphate (Farmer, et al., 1989). Endotoxin levels are measured using the limulus amoebacyte lysate assay. This assay shows that endotoxin levels are usually between 0.6-6 Eu/ml. Finally, sterility is assayed by inoculating thioglycolate broth with LEH, followed by plating on blood agar.

It should be stressed that quality control of LEH solutions is an ongoing research and development area. Many of the standard assay techniques are complicated by the light scattering of liposome solutions and the difficulty in cleanly separating the protein and lipid components of LEH for further analysis. Other examples of manufacturing issues are the control over LEH size distribution for such procedures as sterile filtration. Extraction of LEH lipid components to assay chemical degradation of LEH and long-term stability of LEH components during manufacture, in-vitro storage, and *in-vivo* are ongoing endeavors. An understanding of the encapsulated hemoglobin stability is also an important issue for the large-scale production and quality control of LEH. These concerns have often made an understanding of the effects of LEH *per se*, complicated by manufacturing issues and quality control of LEH solutions.

The preparation of freeze-dried LEH has been previously reported (Rudolph, 1988, Rudolph & Cliff, 1990). The modifications to the described protocol involve substituting the phosphate buffered saline with a buffer that contains the water replacement solute of choice (e.g. sucrose, maltose, trehalose). This is usually done in the first rehydration step followed by processing in the high pressure sheer device. Residual moisture content of lyophilized LEH have shown that preservation in the dry state is optimal at water contents of 1-2%. This is assayed thermogravimetrically or using a colorimetric assay.

Animal Protocol: Balb/c mice were anesthetized using halothane. LEH was infused into either the tail vein or the retroorbital sinus. The animals were then allowed to recover from the anesthetic and given food and water ad libitum. At 2 hours, 24 hours, 1 week, 2 weeks, and 4 weeks post LEH infusion (500 microliters), mice (n=4) were sacrificed and tissues prepared for standard histological examination. Three groups of animals were treated in this study. One group of animals was given a 0.5 ml bolus injection of phosphate buffered saline, one group LEH, and one group LEH that had been lyophilized with 150 mM trehalose and rehydrated just before injection. After sacrifice, organs were fixed in 10% neutral buffered formalin. The following organs were harvested: brain, trachea, esophagus, thyroid gland, salivary gland, mandibular lymph node, adrenal gland, lungs, liver, spleen, kidneys, heart, stomach, duodenum, jejunum, ileum, cecum, colon, rectum, mesenteric lymph node, gonads (ovaries), urinary bladder, femur, bone marrow (femoral), and pancreas. For blood chemistry analysis, blood was removed from the retroorbital sinus and the serum separated by centrifugation. The enzymes serum glutamic oxaloacetic transaminase (SGOT), serum glutamic pyruvic transaminase (SGPT), glutamyl transferase (GGT), and lactate dehydrogenase (LDH) were determined using an Ektochem DT60 analyzer (Kodak, Rochester. NY).

Cellular immune response in mice that have received LEH was measured in mixed leucocyte culture assays. Spleens from animals injected with 0.5 mls LEH were removed after 3 and 7 days. Single cell suspensions were prepared by minimizing the spleen with 5 mls of Click's modified medium without serum. Cells were washed two times before culturing. Leucocytes (5×10^4) (balb/c) were then cocultured in flat-bottom microtiter plates with 1×10^6 heterologous leucocyte stimulator cells (balb/c/J blacks) in Click's medium. Stimulator cells were subjected to a 2000 rad gamma irradiation immediately before coculture. Plates were incubated at 37°C in 5% CO_2 atmosphere for 5 days. At 5 days, 1 uCi [^{3}H]thymidine was added to each well and plates were incubated for another 24 hours. Cells were harvested and cpm determined by liquid scintillation counting. Background was determined from cells cultured without stimulator cells.

Biodistribution of radiolabeled LEH in anesthetized rabbits has been accomplished using a lipophilic chelator of ^{99m}technicium to carry the radiolabel to the LEH interior (Phillips, et al., in press). The biodistribution data has been generated by imaging rabbits under a gamma camera for 2 hours continuously. At 20 hours, the image biodistribution is validated by autopsy and tissue sampling.

RESULTS AND DISCUSSION:

After the initial demonstration of efficacy in total isovolemic exchange transfusion, we began to examine the efficacy of small amounts of LEH in various animal models. One of the first studies which we completed was the hemodynamic consequences of LEH administration in the chronically instrumented conscious rat (Rabinovici, et al., 1989). These studies defined a number of reactions to small amounts of LEH. Again, these results have to be considered in light of the lack of tight quality control measures for LEH preparations. Probably the most significant findings of these studies were the transient thrombocytopenia, leukocytosis, and reduced cardiac output in rats given a 10% normovolemic infusion of LEH (Rabinovici, et al., 1989). These effects diminished 30-60 minutes following the infusion.   Although a mechanistic understanding of these effects is still lacking, one suggestion was that LEH, or alternatively, a contaminant in the LEH solution was causing these effects.   Some of these effects were reduced by changing the source of one of the components of the liposome vehicle, from natural hydrogenated soy lecithin to synthetic distearoyl phosphatidylcholine and by the addition of a platelet activating factor antagonist (Rabinovici, et al., 1990a, Rabinovici, et al. 1990b).   The natural lecithin source contained 2-5% lysolecithin which may have been the cause of some of the observed reactions (Lochner, et al., 1985, Vidaver, et al., 1985, Triarhou & Herndon, 1986, Rabinovici, et al., 1990a, Rabinovici, et al., 1990b).

Biodistribution of radiolabeled LEH in the rabbit has been used to investigate the physiologic traffic pattern of LEH.   These studies show that after a 25% normovolemic infusion of LEH into the ear vein, 50% of the LEH remains in the blood pool at 20 hours. The imaged organ distribution of LEH in the rabbit at 20 hours is shown in Table I.

Table I. *In-vivo* Biodistribution of ^{99m}Tc-LEH in Anesthetized Rabbit, 20 hours post-infusion.

ORGAN	% IMAGED ACTIVITY
Blood	50
Liver	17
Spleen	15
Bladder	3
Kidney	3
Lung	2
Heart	trace
Brain	trace
Urine	trace

This indicates that LEH is primarily removed by reticuloendothelial system organs, principally the liver and spleen. There is no significant disposition of LEH to the kidney which has historically been a sight for hemoglobin toxicity (Bunn, et al, 1969). A more complete description of the biodistribution of LEH in the rabbit and a new procedure for radiolabeling liposomes with technicium will be the subject of forthcoming papers.

The biodistribution studies suggest that pathological consequences of LEH administration could be important, especially in those organs that are primary sites of LEH removal. We have recently completed a preliminary pathology study in mice which were given similar doses to the dose given in the rabbit to generate biodistribution data. The purpose of this preliminary study is to screen a large tissue list to identify possible sites of LEH pathology. All of the animals given LEH solution survived until their sacrifice time (4 weeks was the last time point). Histological examination of the tissues harvested show some lesions in only a few organ systems which may be related to the administration of LEH. In the lung, in early samples (2 hours and 24 hours), vacuoles could be seen in the alveolar capillaries. These vacuoles were 3-7 microns in size and thus could not be equivocally related to the presence of LEH. In addition, the amount of capillary contents may be affected by the degree of exsanguination at sacrifice and necropsy which is not a completely controllable variable. The lung is known to remove large liposomes and since we have 10-30% liposomes of 1-2 microns, the possibility that these lesions may be LEH associated cannot at this time be ruled out. These lesions were not seen in any of the animals sacrificed after 24 hours.

The liver and spleen were the organs dominated by LEH related lesions. At the 2 and 24 hour time points, the liver showed cytoplasmic vacuolization which subsided at the 1 week time point and was not significant at 2 or 4 weeks. The spleen also showed granules in the red pulp, with microscopic visualization of LEH. The spleen handled LEH with similar kinetics, as the 2 and 4 week animals showed reduced lesions. The freeze-dried LEH preparations showed similar pathology to freshly prepared LEH which may indicate that freeze-dried LEH is similar in its biodistribution and handling by the RES. The long-term survival and the normal organ pathology observed at approximately 2 weeks after infusion indicate that at this dose of LEH, there are minimal long-term pathological consequences.

We have also recently begun to address the immunologic consequences of LEH infusion. The purpose of these preliminary studies is to quantify the degree to which leucocytes from mice which have been infused with 0.5 mls of LEH can respond to a challenge from irradiated heterologous leucocytes (see figure 1). Figure 2 shows the tritiated thymidine counts from stimulated leucocyte populations in control animals, and in animals 3 or 7 days post-infusion of LEH. There is no significant difference in the counts observed from control animals or animals exposed to LEH. These results indicate that leucocytes from animals given this dose of LEH are unimpaired in their ability to respond to this stimulus. Future experiments will examine the biochemical response (e.g cytokine production) to LEH by lymphocytes and the effect of LEH on macrophage function.

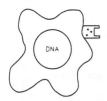

responder leucocyte
Balb/c mouse derived
control and LEH treated 3 & 7 days

stimulator leucocyte
Balb/c/J mouse derived
gamma irradiated

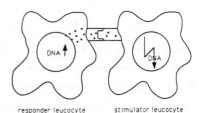

responder leucocyte   stimulator leucocyte

5 days incubation, 24 hours pulsed with 3-H-thymidine

Figure 1. Mixed Leucocyte Reaction (MLR) protocol with leucocytes from mice after injection of LEH.

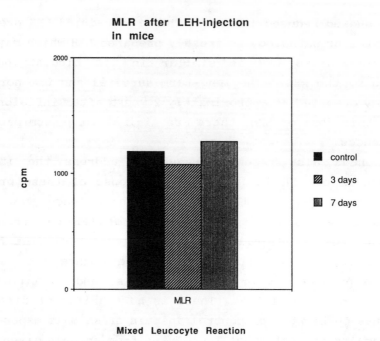

Figure 2. Mixed Leucocyte Reaction assay with splenic leucocytes from mice 3 and 7 days post-injection of LEH.

Preliminary selected blood chemistry analysis of mice given a 25% normovolemic infusion of LEH is shown in figure 3. The two serum transaminase enzymes, SGOT and SGPT are sensitive to organ damage, principally the heart and liver. Their release into circulation is routinely used clinically as indicators of organ injury. The peptidyl transferase enzyme GGT is also very sensitive to liver function. The results indicate that there is little indication of liver damage at the 3, 7, or 14 days post-infusion of LEH. The blood chemistry analysis also demonstrates that the pathological findings of liver accumulation at early time points (2 hours to 1 week) are not manifest in significant impairment of metabolic liver function. It is not clear whether the RES is saturated at the given dose of LEH which may relate to these results. Further experiments into the interaction of LEH with organ systems of the RES are warranted and will be the focus of future experiments.

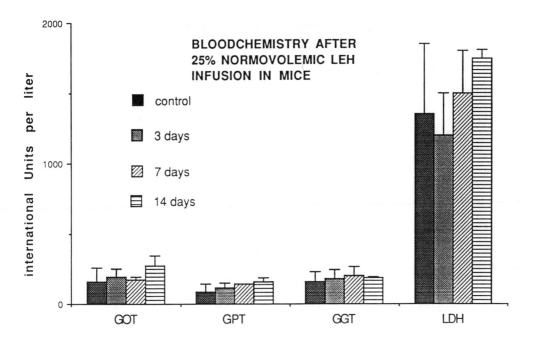

Figure 3. Selected blood chemistry analysis of mice 3, 7, and 14 days post-injection of LEH.

FUTURE DIRECTIONS:

It is our hope that our research efforts will further the development of LEH for clinical use. There are remaining issues concerning quality control and optimization of LEH manufacture. The population size dynamics of LEH is not only a determinant of physiologic traffic pattern, but is important in such manufacturing issues as encapsulation efficiency and sterile filtration methods. We would like to evaluate the use of other hemoglobin sources for encapsulation, particularly recombinant human hemoglobin, as this offers both reduced risks for transmittable antigens and increased availability. It may be that encapsulation of modified hemoglobins (cross-linked or polymerized) may also reduce potential toxicities. Perhaps more importantly, additional efficacy data is needed on LEH, focusing on exchange models of hemorrhagic shock which more closely mimic the clinically relevant application of LEH.

Acknowledgements: The authors wish to thank Ms. Shelley DeLozier for technical editing of the manuscript and the Naval Medical Research and Development Command and the Office of Naval Research for their financial support to A.S.R and W.T.P.

REFERENCES:

Beissinger, R.L., Farmer, M.C., Gossage, J.L. (1986) Trans Am Soc Artif Intern Organs, 32, 58-63
Benesh, R., Yung, S., (1973) Annal Biochem, 55, 245-248
Bunn, F.H., Esham, W.T., Bull, R.W. (1969) J Exp Med, 129, 909-924
Djordjevich, L., Miller, I.F., (1980) J Exp Hematol, 8, 584-592
Djordjevich, L., Pauli, B., Mayoral, J., Ivankovich, A.D., (1982) Anesthesiology, 57, A143
Djordjevich, L., Mayoral, J., Miller, I.F., Inakovich, A.D., (1987) Crit Care Med, 15, 318-323
Farmer, M.C., Rudolph, A.S., Vandegriff, K.D., Hayre, M.D., Bayne, S.A., Johnson, S.A., (1989) In: Blood Substitutes (Marcal Dekker, pp 289-299)
Farmer, M.C., Gaber, B.P., (1987) Methods Enzymol, 149, 184-200
Gaber, B.P., Farmer, M.C., (1984) In: The Red Cell: Sixth Ann Arbor Conference (New York, Alan R. Liss, Inc. pp 179-190)
Gilroy, D., Shaw, C., Parry, E., Odling-Smee, W., (1988) J. Trauma, 28, 1312-1316
Hunt, C.A., Burnette, R.R., (1983) In: Advances in Blood Substitute Research (New York, Alan R. Liss, Inc. pp 59-69)
Hunt, C.A., Burnette, R.R., MacGregor, R.D., Strubbe, A., Lau, D.T., Taylor, N., Kawada, H., (1985) Science, 230, 349-360
Jopski, B., Purkl, Y.U., Jaronix, H.W., Schubert, R., Schmidt, K.H., (1989) Biochim Biophys Acta, 978, 79-84
Ligler, F.S., Stratton, L.P., Rudolph, A.S., (1989) In: The Red Cell: Seventh Ann Arbor Conference (New York, Alan R. Liss, Inc., pp 435-455)
Lochner, A., van-Niekerk, I., Kotye, J.C., (1985) J Theor Biol, 80, 363-376
Miller, I.F., (1981) Chem Eng Commun, 9, 363-370
Phillips, W.T., Timmons, J.H., Klipper, R., Blumhardt, R., Rudolph, A.S., (1990) J Nuc Med, in press
Rabinovici, R., Rudolph, A.S., Feuerstein, G., (1989) Circ Shock, 29, 115-132
Rabinovici, R., Rudolph, A.S., Yue, T.L., Feuerstein, G., (1990a) Circ Shock, 31, 431-445
Rabinovici, R., Rudolph, A.S., Feuerstein, G., (1990b) Circ Shock, 30, 207-219
Rudolph, A.S., (1988) Cryobiology, 25, 277-284
Rudolph, A.S., Cliff, R.O., (1990) Cryobiology, 27, 585-590
Triarhou, L.C., Herndon, R.M., (1986) Arch Neurol, 43 121-125
Tomita, S., Enoki, Y., Yoshida, H. Yasumitsu, Y., (1968) J Nara Med Assoc, 19, 1-41
Vidaver, G.A., Ting, A., Lee, J.W., (1985) J Theor Biol, 115, 27-41

Progress in Membrane Biotechnology
Gomez-Fernandez/Chapman/Packer (eds.)

ROLE OF MEMBRANE COMPONENTS ON ANTHRACYCLINE CYTOTOXICITY

J. M. Gonzalez-Ros, J. M. Canaves, F. Soto, J. Aleu,
A. V. Ferrer-Montiel, and J. A. Ferragut.

Department of Neurochemistry and Institute of Neurosciences,
University of Alicante, 03080 Alicante, Spain.

SUMMARY: The plasma membrane of tumour cells is receiving
increasing attention in regard to cellular multidrug resistance.
We have studied plasma membranes from wild P388 murine leukemia
cells and from stable multidrug resistant sublines with primary
resistance to daunomycin, which do not express P-glycoproteins.
The results obtained with the isolated plasma membranes suggest
that there is a role for certain lipids, namely
phospatidylserine and cholesterol, whose relative abundance is
different in membranes from drug-sensitive or -resistant cells,
in determining (i) the extent of daunomycin binding to the
membrane and (ii) the location of daunomycin within the membrane
bilayer. Furthermore, calorimetric studies on the interaction
between model lipid vesicles with daunomycin and/or verapamil,
the best known resistance-reverting agent, indicate that
verapamil prevents, in a concentration-dependent manner, the
alterations in the phospholipid phase transition expected from
the presence of daunomycin in the bilayer.

Our results suggest that the lipid bilayer of the plasma
membrane could provide appropriate sites where both, the

interaction of antineoplastic drugs and the effects on their
activity of resistance-reverting agents, can be partly
modulated.

THE MULTIDRUG RESISTANCE PHENOTYPE

Multidrug Resistance (MDR) exhibited by tumor cells is
characterized by cross-resistance to a variety of
chemotherapeutic drugs including anthracyclines, Vinca
alkaloids, colchicine and other natural products, thus,
constituting a major clinical problem in the treatment of
cancer. When confronted to those cytotoxic agents, most MDR cell
lines exhibit a reduced intracellular accumulation of the drugs
and a decreased sensitivity to these agents, relative to those
observed in the parental drug-sensitive cell lines (Fine &
Chabner, 1986; Bradley et al., 1988). Development of the MDR
phenotype is related to alterations of the cellular plasma
membrane and, in fact, the observations on a reduced
intracellular drug accumulation have been explained based on the
overexpression of certain membrane glycoproteins (the P-
glycoprotein family). These proteins act as ATP-dependent pumps
that actively eliminate the drugs from the cells (reviewed in
Riordan & Ling, 1985; Bradley et al., 1988; Gottesman & Pastan,
1988; Endicott & Ling, 1989; Ford & Hait, 1990), and constitute
the most solid hypothesis to explain MDR exhibited by cell lines
"in vitro". On the other hand, several hydrophobic cationic
compounds (termed "chemosensitizers"), including well known
calcium channel blockers such as verapamil, calmodulin
inhibitors and others, have been shown to increase the
sensitivity of MDR cells to the cytotoxic drugs and the
intracellular drug accumulation, thus, reverting drug resistance
(Naito & Tsuruo, 1989; see also Ford & Hait, 1990 and references
therein). Furthermore, based on photoaffinity labeling with
verapamil photoactivatable analogs (Safa et al., 1987 and 1988;
Kamiwatari et al., 1989; Bruggermann et al., 1989; Qian & Beck,

1990) and on drug binding studies (Cornwell et al., 1987; Akiyama et al., 1988; Naito & Tsuruo, 1989), it has been suggested that the drug-enhancing activity of verapamil occurs by competition with the antitumor drugs for binding to common sites on the P-glycoproteins. The obvious conclusion emerging from the above comments is that, while overexpression of P-glycoproteins confers drug resistance, the presence of compounds able to compete for drug binding sites on the P-glycoproteins, increases drug sensitivity by preventing an efficient elimination of the drug from the cells. Nonetheless, whether the MDR phenotype occurs and is responsible for drug resistance through a mechanism based on P-glycoproteins in human tumors is still an open question (Ford & Hait, 1990) and in fact, there are reports on MDR cell lines and tumors which do not express P-glycoproteins (Bell et al., 1985; Fojo et al., 1987; McGrath & Center, 1987; Danks et al., 1987; Norris et al., 1989; Cole et al., 1989; Garcia-Segura et al., 1990). Therefore, it should be expected that mechanisms other than those mediated by P-glycoproteins, are also involved in conferring drug resistance to tumor cells and in producing the cellular response to revertant agents. In this context, it should also be noticed that freeze-fracture studies have shown that similarly resistant cell lines expressing (Arsenault et al., 1988) or not (Garcia-Segura et al., 1990) the P-glycoproteins, exhibit similar ultrastructural alterations at their plasma membranes.

In addition to P-glycoproteins, other alterations have been detected in MDR cells, including changes in the expression or in the activity of enzymes involved in the glutathion detoxification pathway (Hamilton et al., 1989), decreased levels of topoisomerase II activity (Sullivan et al., 1989), and many others (Van der Bliek et al., 1986a), and it is possible that some of them could constitute the basis for alternative mechanisms of MDR. However, it is still unknown whether these alterations found in MDR cells have a significant or regulatory role, rather than being simply coincidental or the result of co-expression with P-glycoproteins (Van der Bliek et al., 1986b;

230

Ford & Hait, 1990).

As an alternative (or in addition) to the overexpression of
P-glycoproteins by drug-resistant cells, it has been proposed
that changes in the affinity of the drugs for cellular targets,
could also lead to a reduction in the intracellular steady-state
level of the drugs (Beck et al., 1983). This possibility is
based on the presence of several ionizable groups in many
anticancer drugs, with pK_a's falling within the physiological pH
range. For instance, the anthracyclines are potent antitumor
agents (Arcamone, 1985) bearing an ionizable amino group with a
pK_a between 7.6-8.2 (Skovsgaard & Nissen, 1986). Thus, at
physiological pH, these drugs exist in equilibrium between
neutral and ionized species. The neutral forms are more
permeable through the membranes than the ionized forms, while
the latter have a higher affinity for several cellular targets
(Siegfried et al., 1985). Therefore, the predominance of either
one of these forms at equilibrium should be relevant in
determining cellular drug cytotoxicity and in fact, altering the
degree of dissociation of these drugs has been reported to
greatly influence processes such as the rate of uptake and the
steady-state intracellular level of the anthracyclines
(Skovsgaard & Nissen, 1986). Also, anthracyclines have been
found to accumulate into acidic intracellular compartments
(Sehested et al., 1987), or into artificial lipid vesicles
subjected to a transmembrane pH gradient (interior acidic)
(Mayer et al., 1986). Moreover, a recent theoretical model based
on Ehrlich ascytes cells (Demant et al., 1990), has predicted
that an increase in intracellular pH from 6.95 to 7.4 would be
sufficient to reduce in a 35% the steady-state level of
accumulation of the anthracycline daunomycin. In agreement with
such expectations, we found that increasingly drug-resistant
P388 cells, exhibit an increase in intracellular pH of up to
0.2-0.3 pH units, a decreased intracellular drug accumulation
and have little or no expression of P-glycoproteins (Soto et
al., 1991). However, other murine leukemia cell line, the drug-
resistant L1210 cells, were found not to undergo intracellular

pH change but clearly overexpressed P-glycoproteins. These
observations further emphasize the complexity and variability of
the MDR phenotype and suggest that different cell lines could
have different modes to protect themselves against the presence
of cytotoxic drugs. In this communication we review the
evidence, produced mostly in our laboratory, supporting the
hypothesis that alteration of the lipid components of the plasma
membrane in drug-resistant P388 cells could also provide the
basis for an additional mechanism of defense against the drugs.

ROLE OF MEMBRANE LIPIDS ON DAUNOMYCIN BINDING

Since the report by Tritton & Yee (1982) on the cytotoxic
activity of polymer-immobilized anthracyclines, the plasma
membrane of tumor cells is considered an important cellular
target for these antineoplastics. In an attempt to establish
whether cellular drug resistance is accompanied by alterations
in the extent of anthracycline binding to the plasma membrane,
we used plasma membrane fractions isolated from drug-sensitive
and -resistant P388 cells to perform daunomycin binding studies
using both, steady-state fluorescence anisotropy and
ultracentrifugation procedures to distinguish between free and
membrane-bound drug (Escriba et al., 1990). Our conclusion from
these studies was that plasma membranes from the drug-sensitive
cells bind more daunomycin than those derived from the drug-
resistant cells. Furthermore, drug binding to the membranes was
not affected by either (i) thermal denaturation of membrane
proteins or (ii) proteolytic treatment with trypsin, thus,
suggesting that the protein components of the membrane do not
play a major role in determining neither the extent, nor the
observed resistance-related differences in drug binding. In
support of this, fluorescence resonance energy transfer (FRET)
between the protein tryptophan residues and daunomycin indicates
that, regardless of whether plasma membranes from either cell
line are used in the experiments, the energy transfer process

occurs with a similar efficiency. This further suggests that although membrane proteins may be in close proximity to the incorporated daunomycin, there are no resistance-related proteins markedly involved in the incorporation of the drug.

A comparison between the lipid composition of plasma membranes from drug-sensitive and -resistant P388 cells indicated differences in the relative abundances of certain lipid components, namely, phospatidylserine (PS) and to a much lesser extent, cholesterol (Escriba et al., 1990). To test whether these lipid components could partly be responsible for the observed drug binding, artificial egg phosphatidylcholine (PC) vesicles containing variable amounts of PS and/or cholesterol were used. The results obtained indicated that drug binding is decreased or increased, respectively, as the molar fraction of cholesterol or PS within the liposomes is increased. The similarities between the effects of PS and those observed in the presence of dicetylphosphate (Ferragut et al., 1988), cardiolipin (Nicolay et al., 1984 and 1988) or phosphatidic acid (Nicolay et al., 1984; Henry et al., 1985) in "negatively-charged" liposomes indicate that, in addition to hydrophobic forces (Burke & Tritton, 1985), electrostatic interactions between the drug and the liposome surface are also important in determining total drug binding to membrane sites. The effects observed in the presence of cholesterol could be interpreted as a competition between the sterol and the drug for the occupation of internal sites within the bilayer. This interpretation is consistent with the observation that daunomycin partitioning into membranes occurs at both, membrane surface and lipid-intercalated domains of the bilayer (Henry et al., 1985; Griffin et al., 1986; Ferrer-Montiel et al., 1988; Dupou-Cezanne et al., 1989; see also the Section below). Considering all of the above, the observed differences in drug binding to plasma membranes from drug-sensitive and -resistant P388 cells seem compatible with the observation that plasma membranes from drug-resistant cells contain less PS and more cholesterol than those from drug-sensitive cells. Moreover, liposomes made at PC-PS-cholesterol

ratios resembling those found in the plasma membrane samples,
reproduce almost exactly the differences found in drug binding
in the native membranes (Escriba et al., 1990). Our conclusion
is, therefore, that mainly PS and also cholesterol have an
important role in modulating the binding of daunomycin to the
plasma membrane and in establishing the differences observed
between samples from drug-sensitive and -resistant cells.
Anthracyclines had been previously shown to bind to artificial
lipid vesicles (Burke & Tritton, 1985; Henry et al., 1985;
Nicolay et al., 1984 and 1988) and to native heart mytochondrial
(Griffin et al., 1986) and Torpedo postsynaptic membranes
(Ferrer-Montiel et al., 1988), where the observed binding of the
drug was also mediated by membrane lipids.

ROLE OF MEMBRANE LIPIDS IN DETERMINING THE LOCATION OF
DAUNOMYCIN WITHIN THE MEMBRANE

As indicated in the previous Section, daunomycin partitioning
into artificial lipid vesicles and certain natural membranes
occurs at both, surface and lipid-intercalated domains of the
bilayer. When daunomycin is incorporated into isolated plasma
membranes from P388 murine leukemia cells, the drug also
partitions between "deep" and "surface" membrane domains. Such
domains have been characterized on the basis of (i) FRET between
1,6-Diphenylhexa-1,3,5-triene (DPH) or 1-[4-(trimethylamino)
phenyl]-6-phenylhexa-1,3,5-triene (TMA-DPH) as energy donors,
which are well known in their positioning within the membrane,
and daunomycin as the energy acceptor, and (ii) quenching of the
fluorescence of the membrane-associated drug by the water
soluble quencher iodide (Ferrer-Montiel et al., 1991). Most
interestingly, the distribution of daunomycin between the two
plasma membrane domains is different depending on whether
membranes from drug-sensitive or -resistant cells are used in
the studies. Thus, in membranes from drug-sensitive cells,
daunomycin is preferentially confined to "surface" domains,

while in membranes from drug-resistant cells, the drug
distributes more homogeneously throughout the membrane. These
conclusions are further confirmed by the iodide quenching
experiments, which indicated that the drug incorporated into
membranes from drug-sensitive cells, is significantly more
accessible to the aqueous environment than that incorporated
into membranes from drug-resistant cells. Since drug-membrane
interactions seem important in anthracycline cytotoxicity (see
previous Section), it is conceivable that these observations on
a different membrane distribution of daunomycin, may be related
to the different sensitivity to the drug exhibited by the drug-
sensitive and -resistant P388 cells.

The observed differences in the distribution of daunomycin in
membranes from drug-sensitive and -resistant cells, can partly
be explained by the differences in the relative levels of
certain lipids in the two types of membranes, mainly PS and
cholesterol (see previous Section). We have analyzed how the
presence of these two lipids affects the location of daunomycin,
by using model lipid bilayers made from egg PC supplemented with
PS or cholesterol at different molar fractions. Increasing the
content of the anionic PS into the liposomes, increases the
efficiency of the energy transfer to daunomycin from both energy
donors; the transfer from "surface" domains (explored by TMA-
DPH), being more sensitive to the presence of the acidic
phospholipid than the transfer from the domain explored by DPH.
This is likely the result of an increased incorporation of
daunomycin into the vesicles containing increasing amounts of PS
(see Previous Section) and further suggests the occurrence of
ionic interactions between the negative charge of the
phospholipid polar head group and protonated forms of
daunomycin. On the other hand, increasing the molar ratio of
cholesterol into artificial PC vesicles, reduces the efficiency
of the energy transfer between DPH or TMA-DPH, and daunomycin.
It should be noticed that the effects of both PS and cholesterol
on altering the distribution of the drug incorporated into the
bilayer, become maximal when their molar ratios in the vesicles

are above 20-25%, which are close to the relative concentrations
of these lipids in the native plasma membrane fractions from the
P388 cells. Furthermore, the large efficiencies of FRET between
DPH and daunomycin, along with the binding of the anthracycline
to the lipid components of the plasma membranes from the P388
cells described in the previous Section, lend further support to
the hypothesis that the transport of anthracyclines through
membranes, may occur by passive diffussion (Dalmark & Storm,
1981; Siegfried et al., 1985). In this regard, it could be
speculated that the observed accomodation of a significant
daunomycin population into "deep" plasma membrane domains in the
drug-resistant cells, could result in an easier diffusion of the
drug through these membranes, as compared to that in membranes
from drug-sensitive cells. This seems also consistent with
fluorescence polarization studies using DPH as a probe for
apparent fluidity, which revealed that "deep" domains of plasma
membranes from daunomycin-resistant P388 cells, are in a more
"fluid" state than those observed in plasma membranes from
daunomycin-sensitive cells (unpublished results), and with the
observation that the penetration of anthracyclines into lipid
bilayers is dependent on the molecular packing of the lipid
(Dupou-Cezanne et al., 1989).

ANTAGONISM BETWEEN DAUNOMYCIN AND VERAPAMIL DETECTED IN A LIPID BILAYER

Since the interaction of antitumor drugs with membranes
seemingly play a role in drug cytotoxicity (reviewed in Tritton
& Posada, 1989), and also because the membrane lipid bilayer is
mostly responsible for the interaction with the drugs (Griffin
et al., 1986; Ferrer-Montiel et al., 1988; Escriba et al.,
1990), we have used high sensitivity differential scanning
calorimetry (DSC) and fluorescence depolarization techniques to
study how the presence of daunomycin and/or verapamil, a

resistance-reverting agent, affect the thermotropic behavior of
model dipalmitoylphosphatidylcholine (DPPC) vesicles (Canaves et
al, 1991). In the presence of daunomycin, it is observed that
the DPPC pretransition is completely abolished even at the
lowest daunomycin concentration used, and that the main
transition is progressively broadened as the daunomycin-DPPC
molar ratio is increased, indicating a loss in the cooperative
behavior exhibited by the phospholipid molecules. Parallel to
the observed broadening, the presence of daunomycin shifts the
transition temperature to slightly lower values and reduces most
significantly the magnitude of the enthalpy change associated
with the phospholipid phase transition. Similar experiments
carried out in the presence of verapamil instead of daunomycin,
indicate that verapamil at low concentrations produces changes
in the thermograms comparable to those induced by the presence
of similarly low concentrations of daunomycin. However, further
increasing the verapamil concentration is much less effective
than comparable increases in the concentration of daunomycin, in
producing a decrease in the cooperativity (broadening) or in the
change in heat capacity associated to the phospholipid phase
transition. Nonetheless, when daunomycin and verapamil are
present simultaneously in the DPPC vesicles, it is observed that
verapamil prevents, in a concentration-dependent manner, the
alteration in the phospholipid phase transition expected from
the presence of daunomycin in the bilayer. For instance, in the
presence of a constant concentration of verapamil, increasing
the concentration of added daunomycin does not cause a
significant alteration on the enthalpy change of the lipid
transition, other than that already caused by the presence of
verapamil in the absence of daunomycin. Furthermore, the
decrease on the enthalpy change observed when the DPPC vesicles
are prepared in the presence of daunomycin is progressively
cancelled by the presence of increasing concentrations of
verapamil.

The thermotropic behavior of the vesicles in the presence of
daunomycin, verapamil or both, was also examined by steady-state

fluorescence polarization measurements using DPH as a probe. In agreement with the calorimetric results, the temperature-dependent fluorescence depolarization of DPH also reports that daunomycin produces a downward shift in the transition temperature and a broadening of the transition, while verapamil produces much smaller changes in such parameters. Moreover, when daunomycin and verapamil are present simultaneously, it is observed that verapamil partly prevents the alteration in the phospholipid phase transition expected from the presence of daunomycin in the DPPC bilayer, which is also in agreement with the observations made using DSC.

Our results clearly indicate that the presence of verapamil prevents daunomycin from altering the thermodynamic parameters of the phospholipid phase transition. Under the conditions used in the DSC and fluorescence polarization experiments, complementary drug binding and circular dichroism studies suggest that the observed interference of verapamil in the daunomycin-phospholipid interaction, occurs without a decrease in the amount of daunomycin bound to the lipid bilayer or without the formation of a daunomycin-verapamil complex. Whatever the mechanism might be, this is the first time to our knowledge that verapamil has been shown to interfere on the interaction between daunomycin and a lipid membrane bilayer. Caution should always be exercised in extrapolating the results from studies in simple model liposomes to the much more complex natural membranes. However, it is tempting to speculate that these observations could be an indication that the lipid bilayer constitutes an appropriate locus to partly account for the modulation of daunomycin by verapamil.

CONCLUDING REMARKS

Evidence has been presented to support the notion that alteration in the lipid components of the plasma membrane of P388 leukemia cells is able to modulate (i) the interaction of the anthracycline daunomycin with the membrane and (ii) the location of the drug within the membrane bilayer. Both of these phenomena are likely to bear on the level of intracellular daunomycin that these drug-sensitive or -resistant cells are able to accumulate. The decreased capacity to accumulate antitumor drugs, perhaps the most general feature of drug-resistant cells known to date, could be at least partly explained based on a diminished drug binding to the plasma membrane and on a different ease of diffusion of the drug through the membrane. Additionally, prior to reaching a degree of drug-resistance associated to P-glycoprotein overexpression, the P388 cells undergo a cytoplasmic alkalinization, which should also contribute to decrease the intracellular accumulation of the drug in the moderately drug-resistant cells by simple laws of chemical equilibrium. Furthermore, the finding that verapamil partly prevents the effects of daunomycin on a model lipid bilayer, might perhaps be indicative that verapamil can antagonize the interaction of daunomycin with cellular sites other than those provided by the P-glycoproteins overexpressed in many resistant cells.

Since most drug-resistant tumors which are seen clinically have a low degree of resistance and many of them do not seem to significantly overexpress P-glycoproteins, it is possible that the observations reported here could have some relevance to those clinical situations.

Acknowledgements: This work has been supported by grants PB87-0790 and PB87-0791 from the DGICYT of Spain. J.M.C. and F.S. are recipients of predoctoral fellowships from "Programa de Formacion de Personal Investigador" and "Generalitat Valenciana", respectively. We thank Dr. L. M. Garcia-Segura and Dr. A. Campos for their helpful assistance in the freeze-fracture and flow cytometry experiments, respectively.

REFERENCES

Akiyama, S., Cornwell, M. M., Kuwano, M., Pastan, I., and
  Gottesmann, M. M. (1988) Mol. Pharmacol. <u>33</u>, 144-147.
Arcamone, F. (1985) Cancer Res. <u>45</u>, 5995-5999.
Arsenault, A. L., Ling, V., and Kartner, N. (1988) Biochim.
  Biophys. Acta <u>838</u>, 315-321.
Beck, W. T., Cirtain, M. C., and Leijko, J. L. (1983) Mol.
  Pharmacol. <u>24</u>, 485-492.
Bell, D. R., Gerlach, J. H., Kartner, N., Buick, R. N., and
  Ling, V. (1985) J. Clin. Oncol. <u>3</u>, 311-315.
Bradley, G., Juranka, P. F., and Ling, V. (1988) Biochim.
  Biophys. Acta <u>948</u>, 87-128.
Bruggermann, E. P., Germann, U. A., Gottesman, M. M., and
  Pastan, I. (1989) J. Biol. Chem. <u>264</u>, 15483-15488.
Burke, T. G., and Tritton, T. R. (1985) Biochemistry <u>24</u>, 1768
  -1776.
Canaves, J. M., Ferragut, J. A., and Gonzalez-Ros, J. M. (1991)
  Biochem. J., in the press.
Cole, S. P. C., Downes, H. F., and SlovaK (1989) Br. J. Cancer
  <u>59</u>, 42-46.
Cornwell, M. M., Pastan, I., and Gottesmann, M. M. (1987) J.
  Biol. Chem. <u>262</u>, 2166-2170.
Dalmark, M., and Storm, H. H. (1981) J. Gen. Physiol. <u>78</u>, 349
  -355.
Danks, M. K., Yalowich, J. C., and Beck, W. T. (1987) Cancer
  Res. <u>47</u>, 1297-1301.
Demant, E. J. F., Sehested, M., and Jensen, P. B. (1990)
  Biochim. Biophys. Acta <u>1055</u>, 117-125.
Dupou-Cezanne, L., Sautereau, A. M., and Tocanne, J. F. (1989)
  Eur. J. Biochem. <u>181</u>, 695-702.
Endicott, J. A., and Ling, V. (1989) Annu. Rev. Biochem. <u>58</u>,
  137-171.
Escriba, P. V., Ferrer-Montiel, A. V., Ferragut, J. A., and
  Gonzalez-Ros, J. M. (1990) Biochemistry <u>29</u>, 7275-7282.
Ferragut, J. A., Gonzalez-Ros, J. M., Ferrer-Montiel, A. V., and
  Escriba, P. V. (1988) Ann. N.Y. Acad. Sci. <u>551</u>, 443-445.
Ferrer-Montiel, A. V., Gonzalez-Ros, J. M., and Ferragut, J. A.
  (1988) Biochim. Biophys. Acta <u>937</u>, 379-386.
Ferrer-Montiel, A. V., Gonzalez-Ros, J. M., and Ferragut, J. A.
  (1991), submitted for publication.
Fine, R. L., and Chabner, B. A. (1986) In: Cancer Chemotherapy
  (H. M. Pinedo and B. A. Chabner, Eds), Elsevier, Amsterdam,
  pp. 117-128.
Fojo, A. T., Ueda, K., Slamon, D. J., Poplack, D. G., Gottesman,
  M. M., and Pastan, I. (1987) Proc. Natl. Acad. Sci. USA <u>84</u>,
  265-269.
Ford, J. M., and Hait, W. N. (1990) Pharmacol. Rev. <u>42</u>, 156-192.
Garcia-Segura, L. M., Ferragut, J. A., Ferrer-Montiel, A. V.,
  Escriba, P. V., and Gonzalez-Ros, J. M. (1990) Biochim.
  Biophys. Acta <u>1029</u>, 191-195.
Gottesmann, M. M., and Pastan, I. (1988) J. Biol. Chem. <u>263</u>,

12163-12166.

Griffin, E. A., Vanderkooi, J. M., Maniara, G., and Erecinska, M. (1986) Biochemistry 25, 7875-7880.

Hamilton, T. C., Masuda, H., and Ozols, R. F. (1989) In: Resistance to Antineoplastic Drugs (D. Kessel, Ed), CRC Press Inc., Boca Raton, Florida, pp. 49-61.

Henry, N., Fantine, E. O., Bolard, J., and Garnier-Suillerot, A. (1985) Biochemistry 24, 7085-7092.

Kamiwatari, M., Nagata, Y., Kikuchi, H., Yoshimura, A., Sumizawa, T., Shudo, N., Sakoda, R., Seto, K., and Akiyama, S. (1989) Cancer Res. 49, 3190-3195.

Mayer, L. D., Balley, M. B., and Cullis, P. R. (1986) Biochim. Biophys. Acta 857, 123-126.

McGrath, T., and Center, M. (1987) Biochem. Biophys. Res. Commun. 145, 1171-1176.

Naito, M., and Tsuruo, T. (1989) Cancer Res. 49, 1452-1455.

Nicolay, K., Timmers, R. J. M., Spoelstra, E., Van der Neut, R., Fok, J. J., Huigen, Y. M., Verkleij, A. J., and De Kruijff, B. (1984) Biochim. Biophys. Acta 778, 359-371.

Nicolay, K., Sautereau, A. M., Tocanne, J. F., Brasseur, R., Huart, P., Ruysschaert, J. M., and De Kruijff, B. (1988) Biochim. Biophys. Acta 940, 197-208.

Norris, M. D., Haber, M., King, M., and Davey, R. A. (1989) Biochem. Biophys. Res. Commun. 165, 1435-1441.

Qian, X., and Beck, W. T. (1990) Cancer Res. 50, 1132-1137.

Riordan, J. R., and Ling, V. (1985) Pharmac. Ther. 28, 51-75.

Safa, A. R., Glover, C. J., Sewell, J. L., Meyers, M. B., Biedler, J. L., and Felsted, R.L. (1987) J. Biol. Chem. 262, 7884-7888.

Safa, A. (1988) Proc. Natl. Acad. Sci. USA 85, 7187-7191.

Sehested, M., Skovsgaard, T., van Deurs, B., and Winther -Nielsen, H. (1987) Br. J. Cancer 56, 747-751.

Siegfried, J. M., Burke, T. G., and Tritton, T. R. (1985) Biochem. Pharmacol. 34, 593-598.

Skovsgaard, T., and Nissen, N. I. (1986) In: Membrane Transport of Antineoplastic Agents (I. D. Goldman, Ed), Pergamon Press, Oxford, pp. 195-213.

Soto, F., Planells-Cases, R., Canaves, J. M., Ferrer-Montiel, A.V., Gonzalez-Ros, J. M., and Ferragut, J. A. (1991) submitted for publication.

Sullivan, D. M., Chow, K-C., and Ross, W. E. (1989) In: Resistance to Antineoplastic Drugs (D. Kessel, Ed), CRC Press Inc., Boca Raton, Florida, pp. 281-292.

Tritton, T. R., and Yee, G. (1982) Science 217, 248-250.

Tritton, T. R., and Posada, J. A. (1989) In: Resistance to Antineoplastic Drugs (D. Kessel, Ed), CRC Press, Boca Raton, Florida, pp. 127-140.

Van der Bliek, A. M., Meyers, M. B., Biedler, J. L., Hes, E., and Borst, P. (1986a) EMBO J. 5, 3201-3208.

Van der Bliek, A. M., Van der Velde-Koerts, T., Ling, V., and Borst, P. (1986b) Mol. Cell Biol. 6, 1671-1678.

Progress in Membrane Biotechnology
Gomez-Fernandez/Chapman/Packer (eds.)
© 1991 Birkhäuser Verlag Basel/Switzerland

HYDROPHOBIC PULMONARY SURFACTANT PROTEINS IN MODEL LIPID SYSTEMS

K.M.W. Keough, J. Pérez-Gil, G. Simatos, J. Tucker, K. Nag,
C. Boland, J. Stewart, L. Taylor, S. Taneva,
L.A. Allwood, and M. Morrow

Departments of Biochemistry, Pediatrics and Physics
Memorial University of Newfoundland
St. John's, Newfoundland, Canada A1B 3X9

SUMMARY: Pulmonary surfactant contains two small hydrophobic proteins, SP-B and SP-C, which are important for its physiological actions. SP-C enhances the adsorption of either dimyristoyl- or dipalmitoylphosphatidylcholine (DMPC or DPPC) to the air-water interface, as well as mixtures containing DPPC and either phosphatidylglycerol (PG) or phosphatidylinositol (PI). It is somewhat more effective with DPPC/PI mixtures. SP-C disturbs the packing of the acyl chains, but not the head groups, of disaturated PC. Studies with a model peptide suggest the SP-C may bind calcium. SP-C alters the packing of DPPC in monolayers. The phase of the PC with which SP-C or SP-B is interacted does not influence its structure. SP-C does not substantially alter the order and motion of the acyl chains of DPPC.

Pulmonary surfactant is material in the aqueous lining layer of the small terminal airways and alveoli of the lung that is essential for proper pulmonary function and stability. The material is comprised of about 90% lipids and 10% protein. Subsequent to its secretion from alveolar type II pneumocytes into the aqueous lining layer (hypophase) some of it forms a monolayer at the air-water interface and it substantially reduces the surface tension of the air-water interface. This leads to a reduction in the work necessary to expand the lung, a reduction in the tendency for alveolar collapse, or atelectasis, at low volumes, an increase in lung compliance, dV/dP, and a decrease in the tendency for intraalveolar edema (Clements, 1962).

Absence of surfactant is the cause of respiratory distress syndrome in premature born infants (Avery & Mead, 1959).

In the condition known as adult respiratory distress syndrome

242

surfactant activity is compromised by the presence of interfering
proteins from plasma that have leaked into the alveolar air
spaces (Ashbaugh et al., 1967; Mason, 1985).

Exogenous surfactants are now employed to treat surfactant
respiratory distress syndrome with some degree of success. A
very limited number of attempts to use exogenous surfactants to
alleviate symptoms of adult respiratory distress have as yet been
tried, with less striking success. The surfactants used are
dispersions of natural surfactant lipids, or simple synthetic
mixtures of phospholipids, or lipids and a detergent. There is
considerable interest currently in studying how components such
as the surfactant proteins enhance surfactant activity with the
aim of designing more effective exogenous surfactants.

SURFACTANT PRODUCTION AND FUNCTION

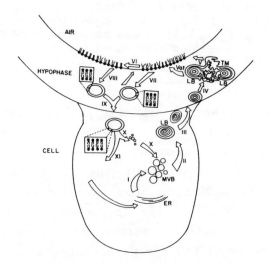

Figure 1. The "life cycle" of surfactant (after Keough (1985),
with permission). Surfactant is synthesized in type II cells and
transferred through multivesicular bodies (I) into lamellar
bodies (II) which are secreted (III) into the hypophase.
Lamellar bodies transform into tubular myelin (IV) which promotes
insertion of lipid into the monolayer (V). Surface refinement
(VI) of the monolayer to enrich it in DPPC occurs. Reuptake (VII
to IX) into the type II cells is followed by reprocessing (X and
XI).

Pulmonary surfactant is synthesized in the type II pneumocytes in the alveolar epithelium. It is stored in the form of condensed bilayer arrays called lamellar bodies, in which form the material is secreted into the aqueous lining layer, or hypophase, of the alveoli.

At least some, if not all, of the material initially secreted in bilayer form transforms into an unusual morphological form called tubular myelin. The presence of tubular myelin is associated with rapid adsorption of lipid into the air-water interface in natural surfactant, and it is generally regarded as the immediate precursor of the monolayer at the surface.

The monolayer which achieves the very high surface pressures characteristic of lung in situ is considered to be enriched in DPPC, through either a process of selective insertion of DPPC, or selective exclusion of non-DPPC components, or both. The selective exclusion interpretation is most commonly accepted.

Surfactant components can be recycled back into the type II cell for repackaging and resecretion.

SURFACTANT COMPOSITION

Pulmonary surfactant is characterized by a relatively unusual composition as shown in Figure 2.

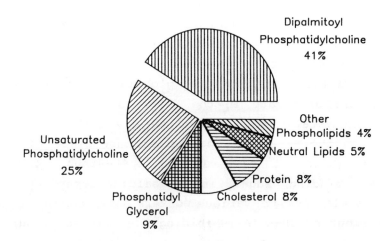

Figure 2. Composition of pulmonary surfactant.

Surfactant Lipids

Surfactant is rich in dipalmitoylphosphatidylcholine (DPPC) which constitutes about 40-45% of the total mass of surfactant. It also contains about 10% acidic phospholipids, with phosphatidylglycerol (PG) being the most abundant of those in most mammals. Monolayers enriched in DPPC are capable of sustaining the high surface pressures (or low surface tensions) associated with lung stability at low lung volume (Hildebran et al., 1979; Hawco et al., 1981). It is considered that the other lipids present in surfactant may aid in the adsorption of lipid to the air-water interface, and in its respreading after it has been compressed into a collapse phase (Clements, 1977, Notter et al., 1983; Fleming & Keough, 1988). Despite the small added benefit to adsorption given by the non-DPPC lipids of surfactant, it has been evident that the rate of adsorption of surfactant lipid from the bulk aqueous phase into the air-water interface is too slow to adequately provide for the rate of adsorption of whole surfactant, or to account for the very fast adsorption properties likely to be required of surfactant in situ.

Surfactant Proteins

Surfactant proteins are named SP-A, -B, -C and -D (Possmayer, 1988; Weaver & Whitsett, 1991; Hawgood & Shiffer, 1991).

SP-A is an octadecamer with monomer molecular weight of 35000. While it is surfactant-associated, it is soluble in aqueous systems. The monomer has a proline-rich portion at its N-terminal and a larger glycosylated globular C-terminal portion. Three monomers associate in the N-terminal region to form a collagen-like "stalk" with the three glycosylated C-terminal portions atop the "stalk." Six trimers associate to form the octadecamer (Voss et al., 1987). SP-A has lectin-like properties, binding sugar in a calcium-dependent fashion (Haagsman et al., 1987). Various workers have found evidence for the involvement of SP-A in inhibiting the release of surfactant from type II cells in culture, enhancing the reuptake of surfactant by type II cells, and in enhancing uptake of bacteria by alveolar macrophages. Of particular interest here is the potential role of SP-A in aiding the transfer of lipid from the hypophase to the air-water interface. The presence of SP-A and calcium promote the formation of tubular myelin and enhance the rate of adsorption of surfactant lipid into the air-water interface. The structure and function of surfactant protein, including SP-A, have been reviewed recently by Weaver and Whitsett (1991) and Hawgood and Shiffer (1991). It has been observed that the combination of SP-A and SP-B can cause the formation of tubular myelin in simple mixtures of DPPC and PG (Suzuki et al., 1989).

SP-D is a collagen-like protein recently identified in surfactant (Persson et al. 1989) whose function remains unknown.

SP-B and SP-C are small hydrophobic proteins, or proteolipids, that are found in the lipid extract of surfactant (Takahashi & Fujiwara, 1986; Possmayer, 1986) Their presence is essential for the very rapid adsorption of lipid from the subphase into the air-water interface. Both proteins are highly conserved among mammalian lung surfactants.

Structural details of SP-B and SP-C are small. SP-B consists of 79 amino acids and contains three regions which may form amphipathic helices (Waring et al., 1989). SP-C, consisting of 35 amino acids, is very hydrophobic with a 23-residue sequence at the C-terminal which contains only hydrophobic amino acids. The more polar N-terminal of SP-C contains palmitoylated vicinal cysteines at positions five and six (Curstedt et al. 1990).

ROLE OF HYDROPHOBIC SURFACTANT PROTEINS

It is well known that fully hydrated liposomes do not rapidly exchange lipid between the bilayer and the air-water interface, and lung surfactant lipid is no exception in this regard. Very rapid adsorption to the interface is, however, a necessary characteristic of surfactant. Relatively recently it became apparent that the hydrophobic proteins found in the lipid extract of lung surfactant were essential for such rapid adsorption (Takahashi & Fujiwara, 1986; Yu & Possmayer, 1986).

Given the normal critical micellar concentration of phospholipids in solution, the movement of lipid from the hypophase into the interface is highly unlikely to occur by diffusion of pure lipids during surfactant adsorption. It is considered that the process likely involves the concerted movement of collections of lipids from the lamellar body bilayers or from tubular myelin into the monolayer. Since the hydrophobic proteins are involved in that concerted movement, one might ask if they can in some way disturb the packing of lipids in a bilayer in a manner different from conventional membrane proteins.

SP-C and adsorption

We have investigated the effect of porcine SP-C on the adsorption rates of simple lipid mixtures containing DPPC and either PG or phosphatidylinositol (PI). Acidic lipids, usually PG but sometimes PI, are considered to be important to surfactant function, possibly especially to the adsorption process. PG is commonly used with DPPC as a component of synthetic surfactants (Morley et al., 1981). PI, while a less common acidic lipid component, seems to lead to a functional surfactant also (Egberls et al., 1987).

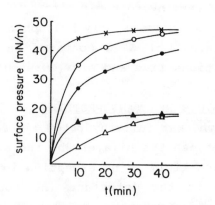

Figure 3. Adsorption into the air-water interface (reflected by increasing surface pressure) of 0.08 mg/mL of lipids plus SP-C. x-x, surfactant lipid extract; ▲-▲, DPPC:PG 7:3 (mole:mole); C: ●-●, + 3% SP-C; Δ-Δ, DPPC:PI 7:3 (mole:mole); o-o, D + 3% SP-C.

Figure 3 summarizes some of the findings on the effect of a fraction enriched in SP-C on the adsorption of mixtures of DPPC plus PG or PI.

It can be seen that SP-C enhanced the spreading from either mixture into the air-water interface, although neither simple mixture was as effective as the whole lipid extract. This could be because other components, such as unsaturated PC and cholesterol, or SP-B, which would be present in the extract also enhance the adsorption rate. Mixtures with PI spread slightly more rapidly than those with PG.

SP-C and bilayer packing

To see if SP-C could cause a special perturbation of lipid packing we have investigated the influence of SP-C on the deuterium magnetic resonance spectra of chain-perdeuterated dimyristoyl PC (DMPC-d_{54}) (Simatos et al., 1990). Figure 4 shows examples of spectra obtained below and above the phase transition of DMPC-d_{54} in the presence of protein.

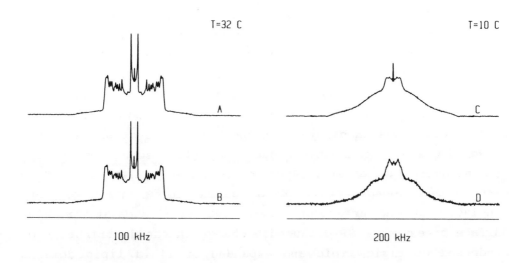

Figure 4: ^{2}H-nmr spectra of dispersions DMPC-d_{54} and DMPC-d_{54} plus 8% SP-C (wt/wt) above (A & B) and below (C & D) the gel to liquid crystal phase transition.

The spectra indicated that SP-C, at low concentrations ($\leq$ 2%) caused the occurrence of two-phase consistence over a small temperature range near the gel to liquid-crystalline phase transition temperature. At high concentration it caused a progressive phase change from an ordered to a disordered structure over a somewhat larger temperature range. SP-C had little effect on chain order in the liquid crystal but it did decrease chain order in the gel. It had no effect on T_1, the spin lattice relaxation time, in either phase, but it decreased the quadropolar echo relaxation time, T_{2e}, in the liquid crystal, and changed its temperature-dependence in the gel. Thus SP-C altered the slower, possibly lateral diffusional, motions of the lipid. No evidence for non-bilayer phase (detection limit of about 5%) was observed. High sensitivity differential scanning calorimetry gave results consistent with these observations. The effect of SP-C on the lipid was, however, not much different than that seen with other membrane spanning proteins.

SP-C in monolayers

Given its extensive hydrophobic character it is possible the SP-C could enter the monolayer formed at the air-water interface, so that a study of its behaviour in monolayers of lipids would be useful in determining its disposition in situ. Also, in a fashion analogous to the use of monolayers to study membrane or bilayer associated properties, the study of SP-C in monolayers can provide useful indications about how it might influence packing of lipids in bilayers in the bulk phase of surfactant.

We have employed the technique of epifluorescence microscopy of monolayers (Nag et al., 1990) to study the interaction of SP-C with DPPC in monolayers. SP-C caused an expansion of DPPC monolayers at low surface pressures, but not so much at very high surface pressures. SP-C caused a change in the distribution of condensed or rigid lipid and expanded or fluid lipid domains during the liquid-expanded to liquid-condensed to solid phases of monolayer compression.

Figure 5 shows that SP-C caused an appearance of more, smaller condensed regions in DPPC monolayers at equivalent areas per molecule for DPPC.

25μ

Figure 5. Images of DPPC monolayers containing 1 mol% of the fluorescent probe NBD-PC and various amounts of SP-C. The dark areas represent condensed domains where the probe has been excluded. The images were obtained at a nominal area/molecule for DPPC of 55 $\mathring{A}^2$/molecule. Left panel, pure DPPC: Right panel, DPPC plus 10% SP-C (by weight).

This effect of SP-C on the packing of DPPC in the monolayer could have occurred if SP-C perturbed the packing of some DPPC molecules as noted above. Then as SP-C concentration increased, the alignment of more molecules of DPPC would be disrupted, causing increasing amounts of fluid lipid to form. This effect would be counteracted by the increasing surface pressure which would tend to produce more condensed lipids. As protein concentration increased, the amount of lipid unaffected by protein would be smaller, and it would be forced by increasing pressure into greater numbers of smaller condensed domains.

Thus SP-C appears to have similar effects in bilayers and

monolayers. Since SP-C did not expand isotherms at very high surface pressures, one must consider the possibility that it is segregated in or excluded from the monolayers under a high degree of compression.

SP-C and calcium

SP-C contains the sequence -Pro Cys Cys Pro- at position 4 through 7. These are the palmitoylated cysteines mentioned above. Depending on their three-dimensional organization, these residues could form a turn which might bind calcium (Ananthanarayanan, et al.). We have observed evidence in 95% acetonitrile-5% 2-chloroethanol for binding of calcium to the 10-residue peptide which is equivalent to the N-terminal of human SP-C with residue one changed from phenylalanine to tryptophan. This suggests the possibility of the existence of a weak binding site for calcium on SP-C. Further studies with palmitoylated peptides and SP-C would be useful in defining the properties of this site.

Synthetic SP-B in bilayers

As has been shown by other workers, we have found that the presence of synthetic human SP-B speeds up the adsorption of DPPC into the air-water interface. We have found that SP-B causes only very small changes in the ^{2}H-nmr spectra of chain perdeuterated DPPC (DPPC-d_{62}).

The ^{2}H-nmr results suggest that SP-B perturbs the chains of DPPC less than does SP-C. Preliminary data from spin-labelled probes in DPPC suggest that SP-B acts closer to the head group than the center of the bilayer. Baatz et al. (1990), using fluorescently labelled lipid probes, have found that SP-B caused ordering of the surface of bilayers of DPPC:PG (7:3) in the gel phase but not in the liquid crystal phase. They found that SP-B did not appear to alter lipid order in the membrane interior.

INFLUENCE OF LIPID PHASE ON THE STRUCTURES OF SP-B AND SP-C

Porcine SP-C and synthetic human SP-B in dispersions of saturated PC have been studied by circular dichroism spectroscopy SP-B was found to have a calculated α-helical content of about 40-45% whereas SP-C had a calculated helix content of about 60-65%. The calculated amount of helix was essentially not much affected by whether or not the lipid was in the gel or liquid-crystalline phase.

SUMMARY

Both the hydrophobic proteins SP-B and SP-C can enhance the adsorption of lipid from bilayers in the bulk phase into the air-water interface. SP-C penetrates into the bilayer and perturbs the packing of acyl chains. Its overall effect on the bilayer is not much different, or more extensive, than that of other proteins which penetrate the lipid bilayer. There was no evidence for induction in DPPC of non-bilayer phase in substantial amounts by SP-C. SP-C can be incorporated into PC monolayers where it also disorders lipid packing. SP-B, on the other hand, has only a limited effect on the lipid acyl chains, but possibly interacts at the level of the lipid head groups. Further work with mixed lipid systems and the hydrophobic proteins will enable more detailed understanding of the actions of the proteins on lipid packing in the bulk phase. Such information will be useful in the design of replacement surfactants for therapeutic use.

Acknowledgements: Supported by the Medical Research Council of Canada and the Newfoundland Lung Association. Synthetic SP-B was from Ross Laboratories and the model peptide for the N-terminal of SP-C was provided by Dr. A. Waring. J.P.-G. was the recipient of a NATO Fellowship.

REFERENCES

Ananthanarayanan, V.S., and Porter, R. (1988) In: Advances in Gene Technology: Protein Engineering and Production (Brew, K. et al., Eds.) IRL Press, Washington, P. 216.
Ashbaugh, M.E., Bigelow, D.B., Petty, T.L., and Levine, B.E. (1967) Lancet 2, 320-323.

252

Avery, M.E., and Mead, J. (1959) Am. J. Dis. Child. 97, 517-523.

Baatz, J.E., Elledge, B., and Whitsett, J.A. (1990) Biochemistry
    29, 6714-6720.
Clements, J.A. (1962) Physiologist 5, 11-28.
Clements, J.A. (1977) Am. Rev. Resp. Dis. 115, 67S-71S.
Curstedt, T., Johansson, J., Persson, P., Eklund, A., Robertson,
    B., Lowenadler, B., Jornvall, H. (1990) Proc. Natl. Acad.
    Sci. USA 87, 2985-2989.
Egberts, J., Beintema-Dabbeldain, A., and de Boers, A. (1987)
    Biochim. Biophys. Acta 919, 90-92.
Fleming, B.D. and Keough, K.M.W. (1988) Chem. Phys. Lipids 49,
    81-86
Haagsman, H.P., Hawgood, S., Sargeant, T., Buckley, D., White,
    R.T., Drickamer, K., and Benson, B.J. (1987) J. Biol. Chem.
    262, 13877-13880.
Hawco, M.W., Davis, P.J. and Keough, K.M.W. (1981)   J. Appl.
    Physiol. 51, 509-515.
Hawgood, S. and Shiffer, K. (1991) An. Rev. Physiol. 53, 575-
    394.
Hildebran, J.N., Goerke, J., and Clements, J.A. (1979) J. Appl.
    Physiol. 47, 604-611.
Keough, K.M.W. (1985) Biochem. Soc. Trans. 13,
Mason, R.J. (1985) West. J. Med. 143, 611-615.
Morley, C.J., Bangham, A.D., Miller, N., and Davis, J.A. (1981)
    Lancet 1, 64-68.
Nag, K., Boland, C., Rich, N.H. and Keough, K.M.W. (1990) Rev.
    Sci. Instrum. 61, 3425-3430.
Notter, R.H., Finkelstein, J.N., and Taubold, R.D. (1983) Chem.
    Phys. Lipids 33, 67-80.
Persson, A., Chang, D., Rust, K., Moxley, M. Longmore, W. and
    Crouch, E. (1989) Biochemistry 28, 6361-6317.
Possmayer, F. (1988) Am. Rev. Respir. Dis. 138, 990-998.
Simatos, G.A., Forward, K.B., Morrow, M.R., and Keough, K.M.W.
    (1990) Biochemistry 29, 5807-5814.
Suzuki, Y., Fujita, Y., and Kogishi, K. (1989) Am. Rev. Resp.
    Dis. 140, 75-81.
Takahashi, A. and Fujiwara (1986) Biochem. Biophys. Res. Commun.
    135, 527-532.
Voss, T., Eistetter, H. & Schafer, K.P. (1988) J. Mol. Biol.
    201, 219-227.
Waring, A., Taeusch, W., Bruni, R., Amirkhanian, J., Fan, B.,
    Stevens, R. and Young, J. (1989) Peptide Research 2, 308-
    313.
Weaver, J.E. and Whitsett, J.A. (1991) Biochem. J. 273, 249-264.
Yu, S.-H., and Possmayer, F. (1986) Biochem. J. 236, 85-89.

Progress in Membrane Biotechnology
Gomez-Fernandez/Chapman/Packer (eds.)
© 1991 Birkhäuser Verlag Basel/Switzerland

SOLUBLE POLYMERIC DRUG CARRIERS : HAEMATOCOMPATIBILITY

R. Duncan, M. Bhakoo, M.-L. Riley and A. Tuboku-Metzger

Cancer Research Campaign's Polymer Controlled Drug Delivery
Group, Department of of Biological Sciences, University of
Keele, Keele, Staffordshire ST5 5BG, England

SUMMARY: Soluble polymeric drug-carriers developed for
parenteral use may be administered intravenously,
intraperitoneally or subcutaneously. Intravenous administration
brings the drug-carrier into immediate contact with blood, and
the cells contained therein. Administration via the other
routes will also ultimately result in transfer of the polymer
conjugate into the circulation. Haematocompatibility of
polymeric drug carriers has not been widely explored, the term
haematocompatibility referring to both the potential
toxicological effects that the carrier may induce on arrival in
the circulation, and the additional possibility that presence in
the bloodstream may interfere with the ability of the drug
carrier to target and/or control the rate of delivery of the
drug payload. Various aspects of haematocompatibility are
discussed and experimental techniques which may prove useful to
define specific parameters are described.

GENERAL CONSIDERATIONS

Before turning attention to factors relating specifically to
haematocompatibility it is important to discuss briefly the
basic concept of a polymeric drug-carrier. For more than a
decade we have been developing soluble synthetic polymer
conjugates for use as drug delivery systems, particularly in
relation to improvement of cancer chemotherapy (Kopecek &
Duncan, 1987). The general rationale for this approach was first
proposed by Ringsdorf (1975) with his theoretically optimised
design of soluble polymeric drug-carriers (or macromolecular
prodrugs). The concept is shown schematically in Fig. 1.

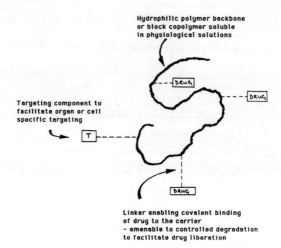

Hydrophilic polymer backbone
or block copolymer soluble
in physiological solutions

Targeting component to
facilitate organ or cell
specific targeting

Linker enabling covalent binding
of drug to the carrier
- amenable to controlled degradation
to facilitate drug liberation

Fig. 1. Schematic diagram showing the theoretical requirements of a soluble polymeric drug carrier

Fixation of an antitumour agent to a hydrophilic polymer automatically changes profoundly both its whole body and cellular pharmacokinetics. Most antitumour agents leave the circulation rapidly, this is certainly the case if they do not become protein bound, whereas conjugation to macromolecular carriers can greatly prolong blood residence time (Seymour et al.,1990), particularly if the carrier used does not target rapidly to a particular organ/cell type. At the cellular level, the macromolecular conjugate does not, unlike most antitumour agents, gain ready access to the cell's interior by membrane penetration, and cellular uptake is limited to the endocytic or lysosomotropic route (as illustrated schematically in Fig. 2), and this factor largely explains the conjugate's extended blood residence time (Duncan, 1987).

Lysosomotropic drug delivery with the aid of polymer drug-carriers affords certain advantages for the development of improved cancer chemotherapy:

＊ Drugs bound covalently to the polymeric carrier via linkages stable in the extracellular environment do not rapidly permeate all cell types, a phenomenon responsible for many dose-limiting toxicities associated with poorly specific antitumour agents.

＊ Careful choice of linkage can optimise rate of drug

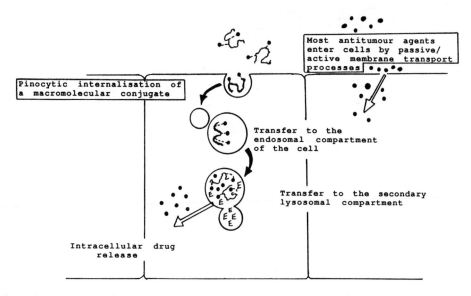

Fig. 2.   Mechanisms of cellular uptake of antitumour agents and macromolecular drug conjugates.

release by lysosomal enzymes, and can thus be used to control the intracellular rate of drug delivery and therefore improve efficacy.

∗ Use of cell specific receptor-mediated pinocytic uptake can be used to improve localisation of the carrier in a target organ or cell type.

To investigate whether the above theory may be useful in practice we have studied extensively drug-carriers based on N-(2-hydroxypropyl)methacrylamide (HPMA) copolymers containing the antitumour agents doxorubicin (Duncan et al.,1989), daunomycin (Duncan et al.,1988) or melphalan (Duncan et al.,1991a) bound to the polymeric background via peptidyl side-chains designed to be stable in the circulation (Rejmanova et al.,1985), but degraded intracellularly by the lysosomal thiol-proteinases (Duncan et al.,1984). In particular two doxorubicin- containing conjugates are being developed for clinical evaluation (Duncan et al.,1991b). These conjugates have a weight average molecular weight (Mw) of approximately 20 000, contain 5-10mol% peptidyl side-chains and approximately 10wt% doxorubicin. The structure of a typical conjugate is shown in Fig. 3.

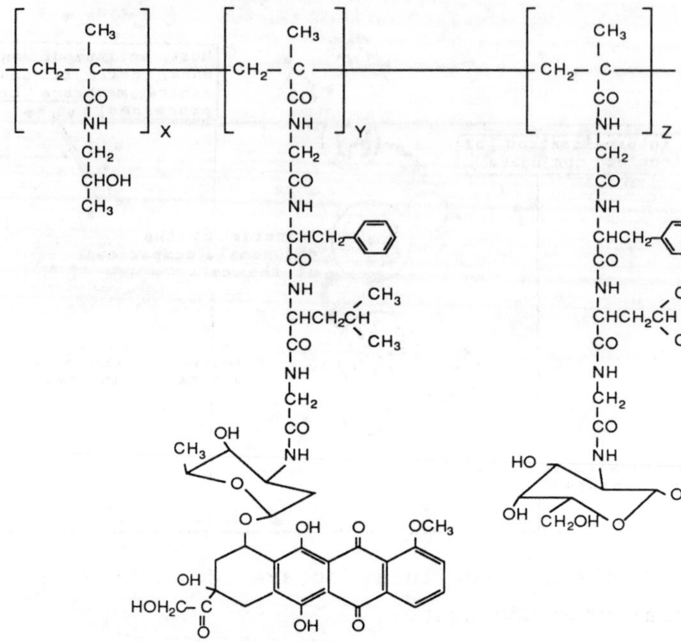

Fig. 3. Structure of HPMA copolymers containing doxorubicin and additionally galactosamine as a targeting moiety. (x~4mol%, y~2mol% and z~4mol%)

HPMA copolymer conjugates containing doxorubicin that are not actively targeted towards tumour cells still display remarkable antitumour activity in vivo against both leukaemic and a variety of solid tumour models including B16 melanoma, P388 and the human colon xenograft LS174T. In part this improved efficacy has been ascribed to the fact that the doxorubicin-polymer conjugates are able to increase drug deposition in solid tumours (when compared with an equidose of free doxorubicin) simply due to their ability to change drug pharmacokinetics (Cassidy et al., 1989). Prolonged plasma half-life of the conjugate together with the known discontinuous endothelial barrier present in the tumour vasculature, and the suggested retention of macromolecules in solid tumours due to poor lymphatic drainage are probably responsible for passive tumour concentration of macromolecules (Maeda & Matsumura, 1989).

Incorporation of targeting residues into HPMA copolymer structure has been used successfully to induce organ-specific targeting (galactosamine to hepatocytes of liver, mannosamine to Kuppfer cells in liver (Duncan et al., 1983)) or enhance cell-specific targeting (melanocyte stimulating hormone to melanoma (O'Hare et al.,1991), antitumour antibodies to colon cancer (Seymour et al.,1991)). Accessability of the conjugate to the target site, ie. ability to extravasate, governs both its blood residence time and its efficiency of localisation in the target organ or tissue. We have shown that polymer modification with galactosamine promotes rapid hepatocyte capture, approximately 80% of dose administered being recovered in the liver within 10min (Duncan et al., 1986).

As the biological properties of HPMA copolymer conjugates have been widely studied, and those containing doxorubicin are being developed for Phase I/II clinical trial, it is important to consider generally their potential interactions in the bloodstream and document their likely haematocompatibility in man. The polymers are both an interesting model system and a delivery system of real clinical relevance.

HAEMATOCOMPATIBILITY

Although many studies have been undertaken to examine the haematocompatibility of polymeric biomaterials (especially those developed as vascular prostheses (NIH Publication, 1985)) the haematocompatibility of soluble polymeric drug-carriers has been largely overlooked. The term haematocompatibility is used here in a very broad sense to describe:

*the possibility that a foreign polymer may induce toxicological/immunological changes that would be harmful to the organism.

*the possibility that the local environment may inactivate the material such that it is unable to perform effectively in its proposed role ie. for site-specific targeting and controlled release of the drug payload.

Both aspects of this definition are equally important. Of course conjugate behaviour which induces toxicity would preclude its use, but additionally any aspects of blood interaction which give rise to inability of the conjugate to deliver drug effectively would also be detrimental, and in the most severe cases would again preclude further development.

First if we consider the toxicological aspects of haematocompatibility. Following entry into the circulation a polymeric drug carrier may interact with either the soluble components (including proteins and antibodies) and/or cellular components, potentially leading to interference with a number of physiological pathways including:

- blood coagulation system

- fibrinolytic system

- kinin system

- complement system

- immune system

Relatively few studies have reported effects of soluble polymeric drug-carriers on these systems, it generally being assumed that polymers with a relatively low LD_{50} must not induce any major changes. However, such an assumption may be dangerous as it is known that polycations such as poly-L-ornithine can activate complement (Asghar,1984), and polyanions, including dextrans substituted with carboxylic and benzylamine sulphonated groups (Mauzac et al., 1985), can reduce complement activation. Simeckova et al (1986) have reported a preliminary study indicating that HPMA copolymers bearing a variety of pendant side-chains inhibit both the alternate and classical complement pathways, but only at concentrations well above the proposed therapeutic range.

The possibility that a soluble polymeric carrier with its pendant drug payload may stimulate an immune response is of considerable concern and this area has been most widely documented (excellently reviewed by Rihova & Riha, 1985). Synthetic polymers may stimulate a humoral (IgG or IgM response), and can also provoke a cellular immune response, in particular polyanions have been shown to activate macrophages. Rihova et al. (1989) have however confirmed that HPMA copolymers bearing doxorubicin do not stimulate a large antibody response.

Although the immune response is by definition initiated by polymer-cell interaction, there are several other cell

interactions that may occur following polymer introduction into
the bloodstream e.g. red blood cell interactions (resulting in
aggregation and/or lysis), platelet interaction, binding to
endothelial cell membranes lining the capillaries and monocyte,
macrophage or lymphocyte interactions leading to lymphokine
release.

As a preliminary screen for polymer-membrane interaction we
have assessed the ability of novel polymers and polymeric drug-
carriers to interact with red blood cells and platelets in vitro
(Bhakoo et al.,1991). A series of routine analytical tests were
developed to monitor red blood cell lysis (haemoglobin release),
aggregation (light scattering measurements at Abs_{600}) or
scanning electron microscopy. Fig. 4 shows the smount of
haemoglobin released during incubation of rat red blood cells
with increasing concentrations of four frequently used soluble
polymers. Whereas the cationic polymers, poly-L-lysine and to a
greater extent polyethyleneimine, cause haemoglobin release, the
accepted plasma expanders dextran and polyvinylpyrrolidone not
surprisingly did not.

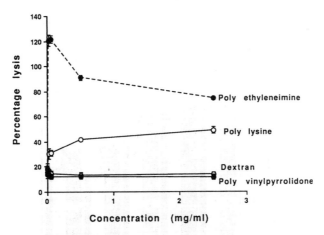

Fig.4 Release of haemoglobin from rat red blood cells following
incubation (shaking at 37°C for 24h in phosphate buffered
saline) with poly ethyleneimine, poly-L-lysine, dextran or poly
vinylpyrrolidone at a range of polymer concentrations. Results
are expressed as a percentage of total haemoglobin released by
Triton-X-100.

Haemoglobin release indicates gross membrane damage, but does

260

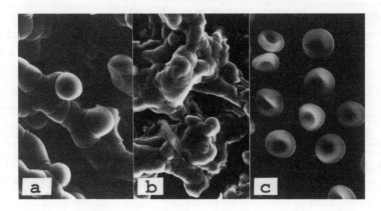

Fig. 5. Scanning electron microscopy of rat red blood cells (mag. x3500) incubated (1h) with poly-L-lysine at concentrations of (a) (0.1mg/ml) or (b) (1.0mg/ml). A time matched control is also shown (c).

not reveal more subtle interactions which could lead to cell aggregation. Scanning electron microscopy provides the opportunity to monitor these events and Fig. 5 shows red blood cell morphology following incubation with poly-L-lysine. At low concentration (0.1mg/ml) the cells begin to aggregate and at higher concentration (1mg/ml) evidence of membrane damage and lysis can be seen.

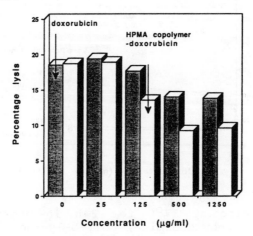

Fig. 6. Release of haemoglobin from rat red blood cells during incubation (shaking, 37°C for 24h) with doxorubicin or HPMA copolymer-doxorubicin. (Data are expressed as a percentage of total released by Triton-X-100)

These observations are consistent with those reported previously by (Katchalsky et al.,1959). The relatively simple standardised assays have proved useful as a preliminary screen to assess many novel polymer-drug conjugates.It can be seen in Fig.6 that HPMA copolymers containing doxorubicin do not enhance red blood cell lysis, and indeed at higher concentrations they appear to protect the cells during the 24h incubation.

DRUG CARRIER PERFORMANCE

Finally, the aspects of haematocompatibiltiy that may reduce drug-carrier performance will be briefly discussed. If we consider a targeted drug-carrier system designed to deliver cancer chemotherapy as mentioned earlier two features are an essential prerequisite for optimal activity:
*ability to target drug either passively or actively to the tumour, ideally with concurrent exclusion of drug from potential sites of toxicity
*ability to ensure controlled release of drug at the target site minimising premature drug offloading in the circulation

Factors effecting macromolecular targeting:
Certain inherent polymer characteristics govern blood residence time and body distribution. Molecular weight of a polymeric carrier regulates; glomerular filtration rate (polymer molecules smaller than approximately 40 000 daltons pass readily through the glomerulus depending on their conformation (Seymour et al.,1987)), rate of extravasation, and phagocytic uptake (high molecular weight polymers are rapidly captured by macrophages (Duncan et al.,1981). Just as seen during opsonisation of particles, interaction of polymeric carriers with soluble or cellular components in the plasma will change apparent molecular weight and physicochemical properties and thus modify its biodistribution. For example, immunoprecipitation by anti-carrier antibodies would prevent the conjugate accessing the target tumour owing to its premature

removal from the circulation. We have shown that the low anti-HPMA copolymer-doxorubicin titre observed in mice does not reduce the therapeutic activity of such conjugates when used to treat leukaemia-bearing animals preimmunised with the same conjugate (Flanagan et al.,1991).

If the drug-carrier is designed to target selectively to tumour cells (by incorporation of antibody, hormone or other targeting group), non-specifc affinity for the endothelial cell wall or other blood cells would reduce the efficiency of the cell-specific targeting system. We have shown that substitution of HPMA copolymers with a hydrophobic drug model (tyrosinamide) led to greatly increased non-specific membrane interaction in vitro above a level of substitution of approximately 10mol% (Duncan et al.,1984) and that a hydrophobic polyamino acid injected intravenously to rats showed considerable liver deposition in vivo (McCormick et al.,1989). Similarly, cationic HPMA copolymers have increased affinity for the liver with more than 50% of the dose administered locating in the liver within 15min (McCormick et al., 1986). All these observations recommend careful preconsideration of these potential interactions for each individual carrier to allow rationale optimistaion of targeting potential. Prudent choice of the carrier, and the degree of drug loading may significantly improve possibilities for tumour-specific delivery.

Factors affecting drug release:

A variety of covalent linkages have been used to attach antitumour agents to polymeric carriers, although amide and ester bonds have been favoured. Rapid hydrolysis in the circulation, or following capture by cells other than the target tumour cells, is likely to reduce the therapeutic index of the conjugate. Careful design of the linker to ensure stability in the circulation can greatly elevate the antitumour capacity of such macromolecular prodrugs. In this context the peptidyl spacers chosen to link doxorubicin to HPMA copolymers proved ideal as proved completely stable in plasma and serum (Rejmanova et al.,1985).

CONCLUSIONS

Soluble polymeric drug-carriers can be designed to increase the therapeutic index of antitumour agents. Consideration of all aspects of haematocompatibilty can assist in the optimisation of such delivery systems, and these considerations fall into two principal areas, polymer interaction with either soluble or cellular blood components. The most important aspects of toxicity and recognition for cell or organ targeting purposes are controlled by cell membrane interactions. There is a fundamental need to increase the number of studies examining such membrane phenomena, using simple in vitro techniques such as; cell receptor binding (target cells, lymphocytes, macrophages, endothelial cells, red blood cells and platelets), model membrane interaction (liposomes, Langmuir trough) and also in vivo investigations particularly relating to receptor specificity and the dose-dependancy of receptor targeting.

Acknowledgements: We are grateful to the British Cancer Research Campaign and the Science and Engineering research Council who support this programme.

REFERENCES

Asghar, S.S. (1984) Pharmacol. Rev. 36, 223

Bhakoo, M., Tuboku-Metzger, A., Riley, M.-L., Hudson, E.A. and Duncan, R. (1991) J. Mater. Sci.: Mater. in Med. in preparation

Cassidy, J.A., Duncan, R., Morrison, G.J., Strohalm, J., Plocova, D., Kopecek, J. and Kaye, S.B. (1989) Biochem. Pharmacol. 38, 875-879.

Duncan, R. (1987) In: Sustained and Controlled Drug Delivery Systems (J.R. Robinson and V.H. Lee, Eds), Marcel Dekker, New York, pp 581-621

Duncan, R., Cable, H.C., Lloyd, J.B., Rejmanova, P. and Kopecek, J. (1984) Makromol. Chem. 184, 1997-2008

Duncan, R., Cable, H.C., Rejmanova, P., Kopecek, J. and Lloyd, J.B. (1984) Biochim. Biophys. Acta 799, 1-8

Duncan, R., Hume, I.C., Kopeckova, P., Ulbrich, K., Strohalm, J. and Kopecek, J. (1989) J. Controlled Rel. 10, 51-63

Duncan, R., Hume, I.C., Yardley, H.J., Flanagan, P.A., Ulbrich, K., Subr, V. and Strohalm, J. (1991a) J. Controlled Rel. in press

Duncan, R., Kopecek, J., Rejmanova, P. and Lloyd, J.B. (1983) Biochem. Biophys. Acta 755, 518-521

Duncan, R., Kopeckova-Rejmanova, P., Strohalm, J., Hume, I.C., Lloyd, J.B. and Kopecek, J. (1988) Brit. J. Cancer 57, 147-156

Duncan, R., Pratten, M.K., Cable, H.C., Ringsdorf, H. and Lloyd, J.B. (1981) Biochem. J. 196, 49-55

Duncan, R., Seymour, L.W., O'Hare, K.B., Flanagan, P.A., Wedge, S., Ulbrich, K., Strohalm, J., Subr, V., Spreafico, F., Grandi, M., Ripamonti, M., Farao, M. and Suarato, A. (1991b) J. Controlled Rel. in press

Duncan, R., Seymour, L.C.W. Scarlett, L., Lloyd, J.B., Rejmanova, P. and Kopecek, J. (1986) Biochim. Biophys. Acta 880, 62-71

Flanagan, P.A., Ulbrich, K. and Duncan, R. (1991) Cancer Letts. in preparation

Kopecek, J. and Duncan, R. (1987) J. Controlled Rel. 6, 315-327

Katchalsky, A., Danon, D. and Nevo, A. (1959) Biochim. Biophys. Acta 33, 120-138

McCormick, L.A., Seymour, L.C.W., Duncan, R. and Kopecek, J. (1986) J. Bioactive and Comp. Polymers 1, 4-19

McCormick, L.A., Sgouras, D. and Duncan, R. (1989) J. Bioactive and Comp. Polymers 4, 252-268

Maeda, H. and Matsumura, Y. (1989) CRC Crit. Rev. Ther. Drug Carr. Sys. 6, 193-210

Mauzac, M., Maillet, F., Jozefonvicz, M. and Kazatchkine, M.D. (1985) Biomaterials 6, 61-63

NIH Publication No. 85-2185 (1985) Guidelines for blood-material interactions

O'Hare, K.B., Duncan, R., Ulbrich, K. and Kopeckova, P. (1991) Brit. J. Cancer 63, Suppl. XIII 23

Rejmanova, P., Kopecek, J., Duncan, R. and Lloyd, J.B. (1985) Biomaterials 6, 45-48

Rihova, B., Bilej, M., Vetvicka, V., Ulbrich, K., Strohalm, J., Kopecek, J. and Duncan, R. (1989) Biomaterials 10, 335-342

Rihova, B. and Riha, I. (1985) CRC Crit. Revs. in Therap. Drug Carrier Syst. 1, 311-374

Ringsdorf, H. (1975) J. Polymer Sci. Polym. Symp. 51, 135-153

Seymour, L.W., Duncan, R., Strohalm, J. and Kopecek, J. (1987) J. Biomed. Mater. Res. 21, 1341-1358

Seymour, L.W., Flanagan, P.A., Al-Shamkhani, A., Subr, V., Ulbrich, K., Cassidy, J.A. and Duncan, R. (1991) Selective Therapeutics, submitted

Seymour, L.W., Ulbrich, K., Strohalm, J., Kopecek, J. and Duncan, R. (1990) Biochem. Pharmacol. 39, 1125-1131

Simeckova, J., Rihova, B., Plocova, D. and Kopecek, J. (1986) J. Bioactive and Compatible Polymers 1, 20-31

Progress in Membrane Biotechnology
Gomez-Fernandez/Chapman/Packer (eds.)
© 1991 Birkhäuser Verlag Basel/Switzerland

POLYHYDROXYALKANOATES, A FAMILY OF
BIODEGRADABLE PLASTICS FROM BACTERIA

Francisco Rodriguez-Valera

Departamento de Genética Molecular y Microbiología, Universidad
de Alicante, Campus de San Juan, Apartado 374, 03080 Alicante

SUMMARY: The accumulation by certain bacteria of poly-ß-
hydroxybutyrate (PHB) as a carbon reserve material has been
known for many years. This polymer is a polyester with the
properties of a thermoplastic similar to polypropylene. In fact,
taking advantage of the inspecificity of the polymerase enzyme,
a whole family of polyesters can be synthesized, with different
mechanical and chemical properties [the so-called poly-ß-
hydroxyalkanoates (PHA)]. These plastics are intrinsically
biodegradable and their use would reduce the damage inflicted
on the environment by petro-chemical plastics due to their
extended lifetime in the environment. Several clinical and
pharmacological applications are also possible and are being
explored. One essential deterrent to the use of bacterial
plastics is their production cost. To reduce it we are
developing a process which uses an extremophile as producer, the
archaebacterium Haloferax mediterranei. The peculiar conditions
in which this organism grows (more than 20% NaCl is present in
the culture medium) prevent contamination by other organisms and
allow the development of extremely large and simple production
facilities. The organism also displays excellent production
parameters and product quality.

PLASTICS AS POLLUTANTS

Plastic remains are probably the worst single problem associated
with solid wastes from industrialized societies. In Spain, for

example, close to 1 million tons of plastic are discarded every year, mostly polyethylene and polypropylene utilized for disposable items such as bags, bottles and assorted containers. One of the greatest virtues of plastics, their durability, becomes a major problem as waste (Pruter, 1987) since they persist in terrestrial and aquatic environments for extremely long periods of time creating severe ecological and aesthetic deterioration. One of the most serious concerns is the pollution of marine environments where several hundred thousands of tonnes of plastic products are discarded every year. The low density of these materials allows them to float and escape burial in the sediment. It has been estimated that more than a million animals are killed every year by choking or entanglement in plastic debris (Bean, 1987). Probably the best solution to the environmental problem created by plastic wastes is the use of biodegradable plastics that, after disposal, are degraded by microbial enzymes, leaving no trace.

POLYHYDROXYALKANOATES

Polyhydroxyalkanoates (PHA), bacterially produced bioplastics, could well become the best biodegradable substitutes of oil-based polymers. They are polymers utilized by bacteria as carbon and energy reserve material and accumulated by them when other essential nutrients are depleted from the medium. The best known, although probably not the most common in nature, is poly-ß-hydroxybutyrate (PHB), an homopolymeric lineal chain of D(-)-ß-hydroxybutyric acid. Described in 1926 by Lemoigne in the Pasteur Institute, PHB is the typical carbon reserve of bacteria, being present in many genera of Gram-positive and Gram-negative bacteria as well as in some archaebacteria (Dawes, 1990). The biochemistry of the polymer synthesis and degradation is well known (Fig. 1) (Dawes and Senior, 1973). The first step is the head to tail condensation of two acetyl-CoA molecules catalyzed by ß-ketothiolase (acetyl-CoA acetyl transferase). The

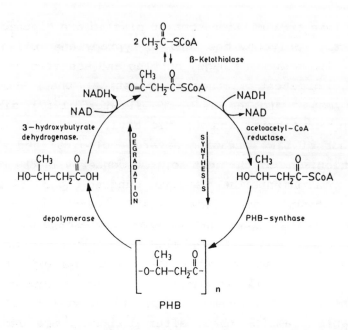

Figure 1. Biochemical pathway of synthesis and degradation of PHB showing the key enzymes.

four carbon intermediary acetoacetyl-CoA undergoes a reduction to ß-hydroxybutyryl-CoA catalyzed by an acetoacetyl-CoA reductase. Finally the four carbon monomers are polymerized by the PHB synthase that establish ester linkage between the acid group of one unit and the hidroxyl group of the next. The polymer is accumulated intracellularly in highly refractile granules of 0.1-0.5 μm diameter surrounded by a thin (2 nm) membrane composed of lipid and protein and to which the activities of both PHB synthase and depolymerase are associated (Figure 2). Intracellular depolymerization occurs via a depolymerase and 3-hydroxybutyrate dehydrogenase that work the pathway in the reverse direction to produce acetyl-CoA. The synthesis of PHB is normally inhibited during balanced growth, apparently due to the inhibition of ß-ketothiolase activity by the high levels of CoASH present in the cells. In nutrient limitation but carbon excess, however, the build up of NADH inhibits citrate synthase, and acetyl-CoA levels rise activating the polymerization pathway

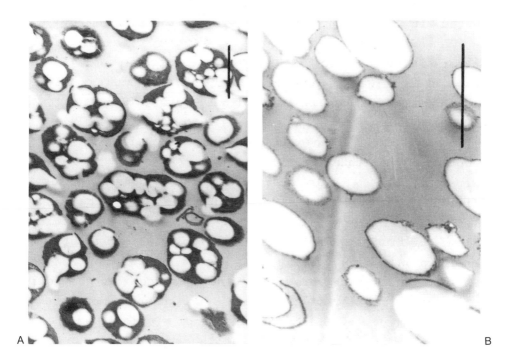

Figure 2. a) Electron micrograph showing cells of <u>Haloferax</u>
              <u>mediterranei</u> grown under optimal conditions for
              PHA accumulation. Large electron transparent
              granules are PHA.
          b) Individual granules, once released from lysed
              cells, showing a conspicuous envelope. Bar=1 um.

(Dawes, 1990). All the enzymes implicated in PHB synthesis have
been cloned and sequenced (Peoples et al, 1987, Peoples and
Sinskey, 1989, Schubert et al, 1988), and in fact PHB granules
have been produced by genetically manipulated <u>E. coli</u> although
the enteric group do not synthesize PHB as carbon storage.

PHA AS THERMOPLASTICS

It is easy to purify PHB from the cells and the resulting
material has the properties of a thermoplastic. During the early
sixties the first patents proposing the use of PHB as a
commercial plastic were filed (Baptist, 1959, Baptist, 1960).

However, it was not until much later that the true potential of the idea was understood, after a polymer containing other monomers in addition to ß-hydroxybutyric acid was isolated from sewage sludge (Wallen and Rohwedder, 1974). These copolymers receive the general denomination of polyhydroxyalkanoates (PHAs) and it is becoming more and more clear that the homopolymers of ß-hydroxybutyrate as an intracellular carbon reserve are the exception rather than the rule.

First of all, it was discovered that ß-hydroxy acids of different chain lengths can be incorporated into the growing polymer, particularly if they are supplied in the culture medium. For example, the soil eubacterium Alcaligenes eutrophus when grown on glucose accumulates the pure PHB polymer, however, when a precursor such as propionic acid is fed into the medium a copolymer containing four and five carbon monomers is synthesized (Holmes, 1985). The amount of propionic acid added to the medium determines the relative proportions of the two monomers. Another microorganism, Pseudomonas oleovorans, can incorporate a whole range of monomers with different chain lengths, thus changing the length of the alkyl pendant group. The nature of the group can also vary widely including cyclic radicals such as phenyl groups when grown with 5-phenylvaleric acid (Fritzsche and Lenz, 1990) or branched chains such as those obtained from 7-methyl octanoate (Figure 3). Finally, the distance between the ester linkage can also vary when 4-hydroxybutyrate or 5-hydroxyvalerate are incorporated into the chain (Doi et al., 1990). Such diversity of polymers derives from an apparent inspecificity of the polymerization system, but the applied interest is the variety of mechanical and chemical properties of the different PHAs that lead to very diverse materials with accordingly diverse applications. However, the ability to obtain radicals of more than two carbons is limited since neither Alcaligenes eutrophus (Anderson et al., 1990) nor Haloferax mediterranei (see below) are capable of producing these polymers, even when precursors are added to the culture medium.

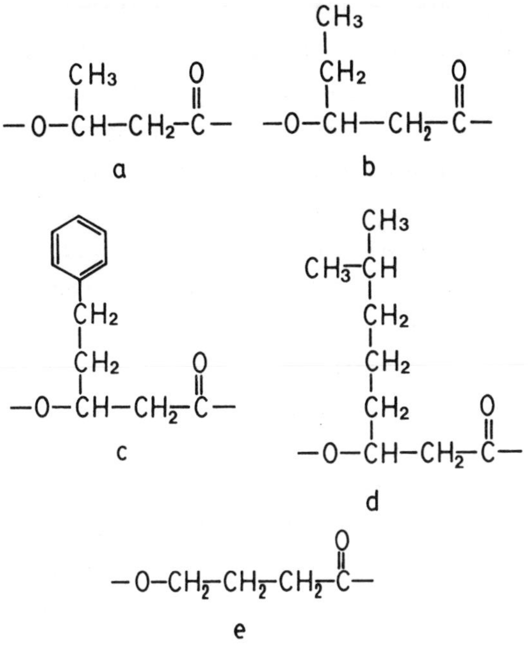

CH3
|
CH3   O   CH2   O
| || | ||
—O—CH—CH2—C— —O—CH—CH2—C—
     a   b

CH3
|
CH3—CH
|
CH2
|
CH2
|
CH2   O
| ||
CH2   —O—CH—CH2—C—
|
—O—CH—CH2—C—
        c   d

O
||
—O—CH2—CH2—CH2—C—
              e

Figure 3. Some examples of monomers incorporated into polyhydroxyalkanoate chains by bacteria: a) 3-hydroxybutyrate, b) 3-hydroxyvalerate, c) 3-hydroxy 5-fenilvalerate, d) 3-hidroxy 7-methyloctanoate d) 4-hydroxybutyrate. a, b and d are incorporated into PHAs produced by different bacteria. c and d have been detected only in PHAs produced by P. oleovorans and related species.

The 3-hydroxybutyrate homopolymer has the properties of a thermoplastic similar to polypropylene. As such the material presents several disadvantages as a commodity plastic. First of all the degree of crystallinity is too high, giving the polymer very low elasticity and high brittleness. On the other hand, its high melting point and severe thermodegradability above the melting point makes the polymer inadequate for thermal processing. Those properties change in a wide range for copolymers. The length of the alkyl pendant groups profoundly affects the crystallinity and macromolecular structure of the polymer. Copolymers of hydroxybutyrate and hydroxyvalerate have much better properties with lower crystallinity level and melting point. Mechanically this results in a decrease in stiffness but an increase in toughness. The melting point is also drastically reduced in copolymers allowing easier processing. Since the amount of hydroxyvalerate units in a copolymer can be varied by simply modifying the composition of the culture medium, a whole range of the above mentioned properties can be obtained, from hard and brittle polymers resembling unplasticized PVC or polyestyrene with the lower range of hydroxyvalerate copolymerization, to soft and tough ones similar to polyethylene with over 20% of HV monomers (Holmes, 1985). Other types of PHA show different properties, for example, copolymers of 3-hydroxybutyrate and 4-hydroxybutyrate possess even lower crystallinity producing extremely elastic polymers (Doi et al, 1990). PHA with long pendant groups possess extremely low melting temperatures and a still lower degree of crystallinity, being elastomers. X-ray diffraction studies of PHA containing C_8 and C_{10} components crystallize differently, they possess a different structure from the 3_1 right-handed helix characteris-tic of HB and HV copolymers, resembling comb-like polymers such as polymethacrylates, poly(vinyl ethers), or poly(vinyl esters) (Gross et al., 1989). Summarizing, bacterial PHAs can be tailored to suit the wide range of applications currently filled by petrochemical plastics, for all uses where biodegradability would be advantageous.

SPECIALITY APPLICATIONS

In addition PHAs can also become a speciality product. For example, by taking advantage of their excellent biocompatibility (they produce an exceptionally mild foreign body response when in contact with animal tissues or body fluids). Moreover PHA's are hydrolyzed in animal tissues (Holland et al., 1987). Those properties make feasible a whole range of pharmaceutical and clinical applications, from retarded drug release (Korsatko et al, 1983) to surgical suture and bone replacements. Another type of application could derive from the stereospecificity of the polymerase system that only incorporates in the polymer the D(-) configuration, therefore, polymerization can be utilized as a stereoisomer purification procedure. D(-) steroisomers are the most commonly found in nature and, therefore, many drugs are only active in this chiral form. Alternatively, the hydrolyzed polymer can be used as a building block in the organic synthesis of such fine chemicals (Holmes , 1985).

BIOPOL: COMMERCIAL PHA

But the exploitation of bacterial PHAs is not just a potential issue. A subsidiary of the British corporation ICI is commercializing copolymers of HB and HV under the trade name Biopol. They obtain the copolymer from the soil eubacterium Alcaligenes eutrophus using a medium with glucose as carbon and energy source and propionic acid as precursor of the HV monomers (Holmes et al., 1981). The organism is cultivated in a fed batch system to improve productivity, phosphorus limitation is the triggering stimulus for the polymer accumulation that reaches up to 80% of the cell dry weight with an overall yield of 0.3 g of polymer per gram of carbon substrate (Byrom, 1987). Biopol can be produced with different percentages of HV modifying the supply of propionic acid. The product can be processed using conventional methods to produce bottles, moulding, fibre and films. In air, Biopol appears to keep indefinitely, even in damp conditions. Underground, however, submerged or mixed with organic rubbish and compost it disintegrates rapidly. Biopol is already being used for cosmetic bottles. However, the present cost of the product (25-30 times the cost of conventional polyethylene) prevents a more extensive market penetration (Pearce, 1990). Besides, with the conventional fermentation technology utilized, to increase the production of Biopol to supply a potential market of hundreds of thousands of tonnes per year would require a huge investment in the production plant.

One way to bypass these limitations would be the use of different organisms as PHA producers. In this regard it is worth noting that the production of PHA by genetically manipulated strains of Escherichia coli has been described (Peoples & Sinskey, 1989). Normal Enterobacteriaceae do not accumulate any form of PHA. By introducing the three genes in A. eutrophus involved in PHA synthesis into E. coli this organism has been reported to accumulate as much as 90% of the cell dry weight as PHB (Slater et al, 1988). Even so, the use of genetically manipulated bacteria for production of low added value materials plants important problems. On the other hand, the use of more remote organisms as recipients for the PHB synthesizing machinery, such as crop plants (Pool, 1989) appears to be a still distant aim.

HALOBACTERIA AS PHA PRODUCERS

We have studied the possibility of utilizing an extremophile for PHA production. Extremophiles are microorganisms that prefer extreme conditions to grow (and often even to survive). The specific advantage derived from using extremophiles in biotechnology resides in their peculiar cultivation conditions that exclude the growth of most or all contaminants. This way the classical industrial microbiology fermenters can be modified, becoming more simple in design and control to the extent of resembling chemical reactors. The scale can be magnified, the process can be carried out by continuous cultivation optimizing production parameters far beyond the levels reached in batch processes. Accidentally we came across an extremely halophilic archaebacteria that is an excellent producer of PHAs: Haloferax mediterranei (Lillo & Rodriguez-Valera, 1990). This organism is a member of the order Halobacteriales, archaebacteria characterized by very high NaCl requirements in their environment (Grant & Larsen, 1989). The halobacteria, as they are often referred to, are aerobic chemoorganotrophs that form large populations in the salt precipitation ponds of solar salterns giving them their typical red hue. The genus <u>Haloferax</u> is characterized by high growth rates (with generation times as short as 3-4 hours) and physiological versatility, being able to utilize many different organic compunds as sole carbon and energy source (Rodriguez-Valera et al, 1983). We first discoverd that <u>H. mediterranei</u> accumulated considerable amounts of PHB when grown in synthetic media with glucose as carbon and energy source (Fernandez-Castillo et al., 1986). Subsequently we have optimized the production of this polymer in batch culture by analysing the effect of different parameters that could influence the production and yield of PHB (Lillo and Rodriguez-Valera, 1990). Of the carbon sources assayed only glucose and starch gave good results. The phosphate concentration greatly influenced the PHB accumulated and the synthesis of the polymer only occured after this nutrient reached a limiting concentration. Phosphate limitation is also the stimulus utilized for PHA production by <u>Alcaligenes eutrophus</u> in the ICI pilot plant (Holmes, 1981). Some production characteristics of the best PHA producers are given in Table I. It is apparent from this data that <u>H. mediterranei</u> ranks second in regard to PHA yield from the carbon source and also the accumulation as percentage of the dry weight. An essential condition for the use of any organism in the production of PHAs is the capability to accumulate copolymers. In the case of <u>H. mediterranei</u> the polymer accumulated when grown on glucose or starch already possesses around 10% of hydroxyvalerate subunits (Rodriguez-Valera & Lillo, 1990). Under the same conditions <u>Alcaligenes eutrophus</u> produces a PHB homopolymer. This behaviour is not unique and has been described for some eubacteria also (Anderson et al, 1990), however, it is fortunate that the percentage of HV subunits found spontaneously in <u>H. mediterranei</u> correspond roughly with the minimum amount required for adequate thermal processing of the material. Additional enrichment in HV subunits can be obtained by addition of propionic acid to the medium, and, even more efficiently, of valeric acid.

Table 1. Comparative production of poly-hydroxyalcanoates by some organisms tested.

Microorganism	Substrate	PHA (% Dry Weight)	Maximum Production (g/l)	Y_{PHA} $(g.g^{-1})$	Reference
Alcaligenes eutrophus NCIB 11599	Glucose	70	–	0.33	[a]
Haloferax mediterranei R4	Starch	67	6.48	0.32	[b]
Haloferax mediterranei R4	Glucose	60	5.13	0.26	[b]
Pseudomonas Sp. K.	Methanol	67	–	0.18	[c]
Pseudomonas cepacia	Fructose	50	2.50	0.12	[d]
Pseudomonas oleovorans	Octanoate	30	0.35	0.21	[e]
Rhodospirillum rubrum	Butyrate	40	1.53	0.08	[f]
Azotobacter beijerinckii	Glucose	70	0.75	0.04	[g]

[a] Holmes, 1981; [b] Lillo and Rodriguez-Valera, 1990;
[c] Suzuki, 1986; [d] Ramsay, 1989; [e] Brandl, 1988;
[f] Brandl, 1989; [g] Dawes, 1973.

CONCLUDING REMARKS

Bacterial PHAs are an original and diverse set of biological polymers that furnish a new tool for material science in several applications that often, but not necessarily always, will take advantage of their biodegradability. Their impact on the global polymer market will depend considerably on the capability of biotechnologists to reduce production costs. Genetic manipulation and the use of novel and better producing organisms represent two ways to achieve this aim.

Acknowledgements: While preparing this review the author was recipient of grants BIO 90-0475 and PTR 89-0003 of the CICYT and PB 87/0792 of the DGICYT (Spanish Ministry of Education and Science). Part of the work described here was funded by grant PBT 86/0011 of the CAICYT. I thank K. Hernandez for secretarial assistance.

REFERENCES

Anderson, A. J, Haywood, G. W., and Dawes, E.A. (1990) Int. J. Biol. Macromol. 12, 102-105.
Baptist, J.N. (1959) U.S.Patent 3'036'959.
Baptist, J.N. (1960) U.S.Patent 3'044'942.
Bean, M.J. (1987) Mar. Poll. Bull. 18, 357-360.
Byrom, D. (1987) Tibtech. 5, 246-250.
Dawes, E.A. (1990) Novel Biodegradable Microbial Polymers. Kluwer Academic Publishers, Dordrecht, The Netherlands.
Dawes, E.A., and Senior, P.J. (1973) Adv. Microb. Physiol. 10, 135-266.
Doi, Y., Segawa, A., and Kunioka, M. (1990) Int. J. Biol. Macromol. 12, 106-111.
Fernández-Castillo, R., Rodriguez-Valera, F., Gonzalez-Ramos, J., and Ruiz-Berraquero, F. (1986) Appl. Environ. Microbiol. 51, 214-216.
Fritzsche, K., Lenz, R.W., and Fuller, R.C. (1990) Makromol. Chem. 191, 1957-1965.
García Lillo, J., and Rodriguez-Valera, F. (1990) Appl. Environ. Microbiol. 56, 2517-2521.
Grant, W.D., and Larsen, H. (1989) In: Bergey's Manual of Systematic Bacteriology (J.T. Staley, M.P. Bryant, N. Pfennig, and J.G. Holt, Eds.), Williams & Wilkins, Baltimore, MD 21202, U.S.A.
Gross, R.A., De Mello, C., Lenz, R.W., Brandl, H. and Fuller, R.C. (1989) Macromolecules 22, 1106-1115.
Holland, S.J., Jolly, A.M., Yasin, M., and Tighe, B.J. (1987)

Biomaterials <u>8</u>, 289-295.

Holmes, P. A. (1985) Phys. Technol. <u>16</u>, 32-36.

Holmes, P.A., Wright, L.F., and Collins, S.H. (1981) Eur. Pat. Appl. # 0 052 459.

Korsatko, W., Wabnegg, B., Tillian, J.M., Braunegg, G., and Lafferty, R.M. (1983) Pharm. Ind. <u>45</u>, 1004-1007.

Peoples, O.P., Masamune, S., Walsch, C. T., and Sinskey, A.J. (1987) J. Biol. Chem. <u>262</u>, 97-102.

Pearce, H. (1990) Scientific European <u>171</u>, 14-17.

Peoples, O.P., and Sinskey, A.J. (1989) J. Biol. Chem. <u>264</u>, 15293-15297.

Pool, R. (1989) Science, <u>245</u>, 1187-1189.

Pruter, A.T. (1987) Mar. Poll. Bull. <u>18</u>, 305-310.

Rodriguez-Valera F., and García Lillo J.A. (1990) In: Novel Biodegradable Microbial Polymers (E.A. Dawes, Ed.) Kluwer Academic Publishers, Dordrecht, The Netherlands.

Rodriguez-Valera, F., Juez, G., and Kushner, (1983) D.J. Systematic and Appl. Microbiol. <u>4</u>: 369-381.

Schubert, P., Steinbüchel, A., and Schlegel, H.G. (1988) J. Bacteriol. <u>170</u>, 5837-5848.

Slater, S.C., Voige, W.H., and Dennis, D.E. (1988) J. Bacteriol. <u>170</u>, 4431-4436.

Wallen, L.L. and Rohwedder, W.K. (1974) Environ. Sci. Technol. <u>8</u>, 576-579.

Progress in Membrane Biotechnology
Gomez-Fernandez/Chapman/Packer (eds.)
© 1991 Birkhäuser Verlag Basel/Switzerland

DEPOSITION OF BIOLOGICAL LIPIDS ON SOLID PLANAR SUBSTRATES BY LANGMUIR BLODGETT TECHNIQUE

G. Puu, I. Gustafson, P.-Å. Ohlsson, G. Olofsson and Å. Sellström

National Defence Research Establishment, Department of NBC Defence, S-901 82 Umeå, Sweden

SUMMARY: With the goal to develop biosensors using membrane proteins as the sensing elements, we have investigated the influence of various factors on successful deposition of phospholipids by Langmuir Blodgett technique to form stable bilayers on planar supports. We use factorial experimental design to screen factors of importance for transfer of the layer separately and for its stability. These experiments were performed on hydrophilic supports (glass, platinum, chromium). The screening included factors such as temperature, pH, ionic composition of the subphase, deposition rate, pretreatment of the support and composition of the lipid mixture. The last-mentioned parameter was also separately studied with a response surface design, in which we varied the proportions of phosphatidylcholine, phosphatidylethanolamine, phosphatidic acid (all as dipalmitoyl derivatives) and cholesterol, to achieve a stable first layer on platinum.

INTRODUCTION

Biosensors of today utilize water-soluble proteins, mainly enzymes and antibodies, as sensing elements. The proteins are immobilized by adsorption, covalent binding or entrapment and are in close contact with a suitable transducer. In nature, recognition and sensing are often membrane-mediated processes. The repertoire of biomolecules useful in biosensor applications would be much wider, if membrane components, e g receptors, could be immobilized with retained biological activity. Different approaches are currently used to solve the problem. Rogers et al (1989) have successfully used a purified nicotinic acetylcholine receptor preparation, adsorbed to a quartz fibre. An optical signal is generated as cholinergic agonists or antagonists bind to the receptor. Although with somewhat altered properties, it is remarkable that the receptor still has the ability to bind ligands, since no amphiphiles, more than the detergent and the lipids remaining from the purification of the receptor, is present. The almost opposite approach is argued for by Buch and Rechnitz (1989), who used the intact chemosensing nerve fibres from the blue crab to detect some excitatory amino acids.

An attractive and exciting alternative to these two approaches is to create a lipid bilayer membrane on the transducer surface and, if appropriate, to incorporate proteins into the lipid film. We advocate that such a system has several advantages, mainly regarding the versatility. Apart from the obvious advantage of having a milieu very much the same as in situ, thus promoting the biological activity of e g receptors, an insulating lipid bilayer would permit the detection of ionic channel activities. We thus have several choices as regards the transducing element of the sensor and can also exploite the amplification property of the biological component. Recognition is, however, a phenomenom not restricted to proteins. Some toxins, such as botulinum, tetanus and cholera, have specific gangliosides as targets and the binding to these lipids is the primary event in the toxic process. The possibility to use such non-proteinous biomolecules as sensing elements in a membrane-based biosensor has, to our knowledge, not been explored. Finally, recent research on lipid bilayers as models for olfaction (Nomura & Kurihara, 1989) and taste

(Hayashi et al., 1990) are suggestive, using the lipid bilayer per se as sensing principle. It also remains to exploite the information that could be hidden in the amplitudes and frequences of the self-sustained electrical oscillations, observed in pure phospholipid bilayer preparations (reviewed in Larter, 1990).

The concept of a membrane-based biosensor is less convincing from a practical point of view. The constraints are mainly linked to various aspects of stability. Reconstituted membranes do not have a long shelf-life. The item of instability is, however, not new in the development of biosensors - the same objection and scepticism were expressed in the early days of development of the water-soluble protein-based sensors. Chemical modifications, covalent bonding between the support and (part of) the lipid monolayer, cross-linking etc might be applicable to circumvent such problems, although any alteration may interfere with the inherent flexible and dynamic properties of the bilayer.

The most attractive method to build up a bilayer of biological lipids on solid supports is the Langmuir-Blodgett (LB) technique. During the last years, this technique has seen a renaissance and is used in many research areas, not the least with applications in electronics, including bioelectronics. While the amphiphilic, naturally occuring phospholipids should lend themselves as suitable for creating LB-films, practical experience is contradictory (reviewed in Roberts, 1990). It is simple to transfer a first layer, but this often adheres so poorly to the support that it is peeled off during the next immersion, i e when the second layer is to be deposited. There are, however, also reports on successful multilayer depositions. We decided to systematically investigate the experimental conditions necessary for obtaining, reproducibly, a well-adhered monolayer on hydrophilic supports (Sellström et al 1991). There are many variables which might effect both the deposition and the stability of the monolayer. The systematic approach is met by using factorial design. We measure the strength of adhesion directly with the LB equipment. The extent by which the deposited lipid monolayer leaves the support when passing it through the air-water interface, with a lipid layer kept at low, constant surface pressure, is recorded. This test ("stripping") was chosen not only because it is simple and gives quantitative information but also because we find it relevant for the applications we aim at. The deposited film must remain intact when taken through different treatments such as transfers between aqueous solutions used in analytical work.

MATERIALS AND METHODS

The Langmuir-Blodgett system was from KSV Instruments Ltd, Helsinki, Finland, and has a trough with a surface area of 15 * 54 cm. The lipid or lipid mixture was dissolved in chloroform. The lipid film was compressed at a constant speed, 5 mN/m/min. Deposition was always at a surface pressure of 45 mN/m. We use three types of hydrophilic substrates - glass, chromium and platinum, starting from microscopic glass slides, 3.5 * 1 cm. The metal-plated substrates were prepared by thermal vacuum evaporation of the metals. Dipping was made to cover 2 cm of the substrate.
We varied, in the screening experiments, many variables and the details are given in Sellström et al., 1991. When studying an optimal lipid composition for deposition on platinum supports, we had 100 μM CaCl$_2$, pH 5, as subphase, dipping rate was 5 mm/min and temperature 25° C.

Test of adhesion: The support was, after transfer of a monolayer (first upstroke), allowed to dry in air for 15 minutes. The surface pressure of the lipid film in the trough was reduced to 5 mN/m and left to stabilize in the meantime. We then immersed again but now at an increased speed, 20 mm/min, and registered the transfer ratio, which could be a negative (lipid left the support) or positive (more lipid was deposited on the support) value.

Experimental design. Screening: A fractional factorial design (Box et al., 1978), with variables at two levels, was applied. The experiments were run in a randomized order. The coefficient for each variable was calculated using multiple linear regression.

Experimental design. Response surface analysis: A central composite design (Box et al., 1978) was chosen to study the effect of lipid composition on the stability of the deposited layer on platinum. We thus varied the proportions of DPPC, DPPE, DPPA and cholesterol in a systematic way. As we knew from the screening experiments that DPPA adheres, under proper conditions, very well to platinum, we argued that this lipid could serve as a reference. We thus handle the quotas DPPC/DPPA, DPPE/DPPA and Chol/DPPA as variables, each at five levels. With three variables, the number of experiments are reduced to 20. All experiments were, however, run in duplicates. The coefficients for the variables were calculated using multiple regression.

RESULTS
Screening of variables of importance for transfer and adhesion: In the first screening experiment we tested 12 variables which could be of importance for the transfer and for the adhesion of the monolayer to the support. Many of the variables influence the packing properties of the lipid film (eg. ionic composition of the subphase, choice of lipid), while others affect the surface properties of the support (eg. type of support and cleaning procedures) or are concerned with the transfer process or the composition of the subphase. The results suggested some variables to be further studied, while

others could be set at a fixed level, either because the variable was without or had a clearcut effect. The following set of screening experiments were made on three types of substrate - glass, chromium and platinum. For the first two, we studied six variables, for platinum five. Cleaning procedures, composition of the subphase, dipping rate during transfer and choice of lipid, each at two levels, were the variables investigated. The lipid compositions compared were 100 % DPPA and a mixture of DPPC (60 mol%), DPPA (15 mol%) and cholesterol (25 mol%).

Although the "stripping" test, i e the adhesion test, is the important one for this work, we always also analyse the effects of the variables on transfer ratio in the deposition step. As we use hydrophilic substrates, one should expect the first deposition during the first up-stroke. This was also the case. We always got a transfer ratio of about 1 at this step. Some variables, such as lipid composition, had statistically significant effects on this value, but from a practical point of view this contribution is less interesting. We also noticed a partial deposition of lipid during the first down-stroke, i e the first immersion of the substrate through the interface in the trough. This was observed especially with the chromium substrate, while platinum had the least tendency, of the supports we used. For both these metal-plated substrates we found that the choice of lipid was important, DPPA giving a higher transfer ratio than the lipid mixture.

In the adhesion test we found that the lipid monolayer was always peeled off, when glass was used as support. None of the variables could significantly alter the amount of lipid leaving the support. We thus conclude that glass is less suitable as support.

The lipid film adhered much better to the metal-plated substrates. We never noticed a complete stripping, and the observed values were both in the negative and in the positive range. The average transfer value in the adhesion test for chromium was satisfactory low, - 0.03. A most significant variable was the choice of lipid, DPPA being much better than the lipid mixture, with a coefficient of 0.34. For platinum we found a more negative average transfer value, -0.12. The choice of lipid, DPPA being superior, was again of great importance. More surprising was that the cleaning of the support was even more important. When platinum was electrochemically cleaned, with a cyclic voltammetry procedure, we found that the lipid film was much more easily peeled off than when the support was only chemically cleaned.

Response surface analysis for optimizing lipid composition: In this experimental series, we used only platinum as support. This was handled as the results in the previous series indicated, and all variables except lipid composition were kept at constant values, according to the results obtained.

While DPPA has adhesion properties that makes it attractive for LB applications, it is less convincing as model, constituting only about 1 % in natural membranes. Is it possible to find mixtures of more abundant lipids, which have the same good adhesion properties? We accordingly included phosphatidylcholine, phosphatidylethanolamine, phosphatidic acid, all as dipalmitoyl derivatives, and cholesterol in the study. When the phospholipids were investigated separately, we found that all were easily transferred to platinum supports and that not only DPPA but also DPPE showed good adhesion, having positive transfer ratios in the "stripping" test. The DPPC-film was on the other hand peeled off to 50-70 %. The challenge was thus to find lipid compositions, with a reasonable proportion of DPPC but still resistant in the adhesion test.

With four lipids, the number of combinations is huge and the experiments would not be manageable, if a traditional "one variable at the time" strategy was used. Furthermore, as interactions between the variables are expected, the results would be difficult to interpret. We thus chose to work with a factorial design also for this part of the study, and used a central composite design (Box et al.,1978). To further reduce the number of experiments, we argued that, as DPPA has the wanted properties as regards adhesion, we could use this lipid as reference and have the quotas DPPC/DPPA, DPPE/DPPA and Chol/DPPA as variables. We also permitted the fraction of DPPA to be greater than is the case in natural membranes.

The variables and the domains used are indicated in Table I. The value of α was set to 1.623. The mole fraction of DPPA was varied between 13 and 36 %, of DPPC and DPPE between 0 and 60 % and of cholesterol between 0 and 28 %.

Table I. Variables and domains for response surface analysis

Variable	LEVELS				
	$-\alpha$	-1	0	$+1$	$+\alpha$
DPPC/DPPA	0	0.76	1.88	3.00	3.76
DPPE/DPPA	0	0.76	1.88	3.00	3.76
Chol/DPPA	0	0.25	0.63	1.00	1.25

This design comprised 20 experiments, which were run in duplicate. We found a good agreement in the duplicates. The transfer ratios in the adhesion test included both negative and positive values, ranging from - 0.87 to + 0.54, while there was almost no variation in the deposition step.

The results from the regression analysis is shown in Table II.

Table II Regression analysis on stripping (S)

Term		Coeff	P
Constant	β_0	-0.22	0.08
DPPC/DPPA	β_1	-0.15	0.07
Chol/DPPA	β_2	-0.06	0.23
DPPE/DPPA	β_3	-0.05	0.30
$(DPPC/DPPA)^2$	β_{11}	0.17	0.00
$(DPPC/DPPA)*(Chol/DPPA)$	β_{12}	-0.01	0.87
$(DPPC/DPPA)*(DPPE/DPPA)$	β_{13}	0.07	0.28
$(Chol/DPPA)^2$	β_{22}	-0.08	0.12
$(Chol/DPPA)*(DPPE/DPPA)$	β_{23}	-0.03	0.66
$(DPPE/DPPA)^2$	β_{33}	0.05	0.33

The coefficients $\beta_0 \ldots \beta_{33}$ represents the estimates of the effects of the variables in the model $S = \beta_0 + \beta_1 x_1 + \ldots + \beta_{12} x_1 x_2 + \ldots + \beta_{33} x_3^2$. The probability value is obtained with t-test and gives the probability that the coefficient is equal to zero. The calculations have been performed with the factors scaled in accordance to Table I.

The statistical analysis, which implies that the response can be described by a second-order model, showed that the ratio DPPE/DPPA was unimportant for the response obtained in the adhesion step. DPPE is thus exchangeable with DPPA, within the concentration domains used. Variables with highly significant effects on "stripping" were the ratio DPPC/DPPA, with negative coefficient, and its quadratic form, $(DPPC/DPPA)^2$, which had a positive coefficient. Less significant was the ratio Chol/DPPA and its quadratic form, which both had negative coefficients.

The model obtained should, if correct, have a predictive value. We could indeed verify the model in control experiments. The agreement between predicted and obtained "stripping" values were better when the coefficients for all variables were included than when only the statistically significant variables were taken into account. As we are interested in the further use of the supported lipid monolayer, we paid special attention to different lipid combinations predicted to result in a good adhesion. One such combination is DPPC:19; DPPE:30; DPPA:32; Chol:19 mol%, which has a rather high (0.3) positive value in the adhesion test and still contains a reasonable proportion of phosphatidylcholine.

DISCUSSION

Phospholipid bilayer formation on solid supports with Langmuir-Blodgett technique differs from and is more complex than formation of other model membranes, such as liposomes. The process takes place at an air-water interface, not in solution. The use of a solid support per se needs special consideration. This should not be looked upon as a passive substratum but as an interacting part in the process. It is known since Blodgett's days that the formation of the first layer is different from the formation of the following ones. Furthermore, there is a disproportion between the dimension of the support to be covered and the thickness of lipid bilayer, and even small irregularities in the support might be of a considerable size in relation to bilayer. During our work with the LB-technique, we have been convinced that great caution in the fabrication and handling is decisive for obtaining reproducible transfer and "stripping" data.

When a monolayer is transferred to the support, an aqueous film is inevitably codeposited, between the support and the lipid film. We believe that the quantity and the composition of this aqueous interspace are decisive for the stability of the film. Before doing the adhesion test, we always let the lipid-covered support dry for about 15 minutes, but it seems reasonable to assume that some water is still present. The amount of water withdrawn and the amount evaporated during the drying procedure should be dependent on the hydrophilicity of the support, a highly hydrophilic surface binding more water more firmly than a less hydrophilic surface. Thus, although a highly hydrophilic surface also should be favourable for a tight binding to the charged headgroups of the lipids, this positive effect might be concealed by the presence of a relatively thick aqueous interspace. The two most hydrophilic substrates used in this study - glass and electrochemically cleaned platinum- were inferior to other, less hydrophilic supports in the adhesion test.

The results from the study with different lipid compositions can also be interpreted from this point of view. Among the lipids used, phosphatidylcholine is much more hydrated than phosphatidic acid and phosphatidylethanolamine, while cholesterol has the least tendency to bind water. Other factors, such as packing parameters and the ability of the negatively

charged phospholipids to bind calcium ions, certainly contribute to the difference in strength of adhesion. The
beneficial effect of having some cholesterol present in the
lipid mixtures became apparent in our experiments to verify
the model obtained in the response surface analysis. This effect could be due to the condensing effect of cholesterol,
resulting in a more densely packed lipid film and thus more of
ionic binding between the lipids and the support and also
stronger hydrophobic interactions between the acyl chains. It
is also possible that the main effect is coupled, directly and
indirectly, to hydration. The direct effect is due to the poor
water-binding properties of cholesterol, while an indirect effect could be due to an interaction between the 3-β-OH group
in cholesterol and one of the carbonyl groups in the glycerol
moiety of phosphatidylcholine (Finean, 1990), which we suggest
results in a reduced hydration of this phospholipid. It might
be more than coincidence by chance that our favourite lipid
mixture contains equimolar amounts of cholesterol and phosphatidylcholine.

CONCLUSION: It is possible, by careful choices of support, of
cleaning and handling of the support, of environmental factors
and of lipids, to find optimal conditions for deposition of a
first, stable monolayer of biologically important lipids. The
support and the lipids are the most important factors. The
factorial experimental design has proven useful and attractive
for our systematic approach to develop model membranes on
solid supports.

REFERENCES

Box, G.E.P., Hunter, W.G. and Hunter, J.S. (1978) Statistics
    for experimenters. An introduction to design, data analysis
    and model building. John Wiley and Sons, New York.
Buch, R.M. and Rechnitz, G.A. (1989) Anal. Lett. 22, 2685-
    2702.
Finean, J.B. (1990) Chem. Phys. Lipids 54, 147-156.

Hayashi, K., Yamanaka, Y., Toko, K and Yamafuji, K. (1990) Sensors and Actuators B, 2, 205-213.

Larter, R. (1990) Chem. Rev. 90, 355-381.

Nomura, T. and Kurihara, K. (1989) Biochim. Biophys. Acta 1005, 260-264.

Roberts, G. (ed) (1990) Langmuir-Blodgett films. Plenum Press, New York.

Rogers, K.R., Valdes, J.J. and Eldefrawi, M.E. (1989) Anal. Biochem. 182, 353-359.

Sellström, Å., Gustafson, I., Ohlsson, P.-Å., Olofsson, G. and Puu, G. (1991) Submitted to Biochim. Biophys. Acta.

Progress in Membrane Biotechnology
Gomez-Fernandez/Chapman/Packer (eds.)
© 1991 Birkhäuser Verlag Basel/Switzerland

"IMMOBILIZATION OF HEPATIC MICROSOMAL CYTOCHROME P450 FROM PHENOBARBITAL TREATED RATS IN HOLLOW FIBER BIOREACTORS"

P. Fernandez-Salguero [*], A.W. Bunch [§] and C. Gutierrez-Merino [*]

(*) Depto. de Bioquimica y Biologia Molecular, Facultad de Ciencias, Universidad de Extremadura. 06080-Badajoz, Spain.

(§) Biological Laboratory, University of Kent at Canterbury, Kent CT2 7NJ. England. U.K.

SUMMARY: Rat liver microsomes can be immobilized in hollow fiber bioreactors under mild physical chemical conditions. The immobilized system shows a cytochrome P450 activity of about 20% of that of isolated microsomes, and retain more than 90% of the initial activity upon incubation for 24 h at 25°C, or 3-4 weeks at -20°C. Thus, these immobilized systems appears to be suitable for technological applications, such as removal of xenobiotics from biological samples or from the blood of mammals.

INTRODUCTION

The cytochrome P450 monooxygenase system is the most relevant membrane complex involved in the detoxification of many lipophilic compounds, such as xenobiotics (herbicides, pesticides, drugs and carcinogens among them) and steroid hormones (Coulson et al., 1984). Two proteins are mainly responsible of the detoxification process: NADPH cytochrome P450 reductase and cytochrome P450. The first one reduces the cytochrome P450 by a two electron transfer mechanism (White et al., 1980). Eventually, the second electron may be transferred by the cytochrome b_5 (Pompon, 1987). Cytochrome P450 modifies the xenobiotic (increasing its solubility in water) by activation with oxygen through reactions like N-, O-, and S-dealkylations, N-oxidations, hydroxylations and epoxydations (Nebert and Gonzalez, 1987).

On the other hand, chronic treatments with xenobiotics can selectively induce specific isoenzymes of this system (Thomas et al., 1987). Namely, it is well known that the chronic treatment of rats with phenobarbital or polychlorinated pesticides produces a large increase in the microsomal content of cytochrome P450 isoenzymes (Åström et al., 1985; Waxman and Walsh, 1982). In good agreement with earlier findings, upon chronic treatment with phenobarbital we have observed an increased content of microsomal cytochrome P450, as well as a 4-5 fold increase of its deethylase activity of 7-ethoxycoumarin.

Hollow fiber bioreactors have been successfully used in the immobilization of cells and enzymes (Bunch, 1988; Estrada-Diaz et al., 1989).

In this communication, we report the immobilization of the cytochrome P450 of hepatic microsomes in hollow fiber bioreactors, and briefly discuss some of the basic properties and advantages of these bioreactors.

MATERIALS AND METHODS

The induction of rat hepatic cytochrome P450 by phenobarbital has been made by intraperitoneal injection during 4 days at a dose of 65 mg/Kg body weight/day (the average body weight was 200-250 g). The rats were killed 24 h after the last dose and liver microsomes were prepared following Lu et al (1972). Protein concentration in microsomes was measured by the method of Lowry (1951). The amount of protein immobilized in hollow fibers has been determined by the method of Lowry as well, using small pieces of chopped fibers previously incubated at 100°C during 5 min in the presence of 3 M sodium hydroxide. The cytochrome P450 content of microsomal preparations was determined by the method of Omura and Sato (1964), and total heme and cytochrome b_5 contents were determined as indicated in Tredger at al. (1984).

The NADPH cytochrome P450 reductase activity was determined with cytochrome c as exogenous electron acceptor, using an extinction coefficient of 21 $mM^{-1} \cdot cm^{-1}$ at 550 nm for the difference absorption spectra between oxidized and reduced protein. All the spectrophotometric measurements have been done with a Hewlett Packard 8451A diode array or with a Kontron, mod. Uvikon 810, spectrophotometers, in 100 mM phosphate buffer (pH 7.2 at 25°C).

Cytochrome P450 activity using 7-ethoxycoumarin as substrate has been measured from the change in the intensity of the emission of fluorescence at 460 nm (excitation wavelenght 360 nm) produced by the product, 7-hydroxycoumarin. The change in fluorescence intensity was converted in enzymatic activity by calibration of the signal with known amounts of 7-hydroxycoumarin.

Scanning electron microscopy has been done on small fragments of chopped fibers dryed with acetone, and coated with gold in a Jeol apparatus. The coated samples were observed in a Jeol electron microscope, mod. T-100.

Immobilization in acrylamide gels. Acrylamide (10-30%) has been polymerized in 100 mM phosphate buffer containing methylene-bisacrylamide (5% with respect to acrylamide), 0.5% v/v TEMED and 0.1% w/v ammonium persulfate. This solution is dispersed, under magnetic stirring, in 6 volumes of toluene:chloroform (73:27) containing 0.25% v/v Tween 20 as stabilizer. After 30 min the polymerization is completed, and the organic solvents are washed away with ca. 200 fold volumes of distilled water by filtration with a water pump. Once prepared, the gel has been activated with glutaraldehyde by incubation under mild stirring conditions during 18-22 h in 100 mM potassium phosphate (pH 7.2). The excess of glutaraldehyde was removed by washing with 100 volumes of distilled water in a water pump. Covalent binding of microsomes to the gel was achieved by 12-15 h incubation at room temperature under mild stirring. Usually the protein concentration was fixed to 0.5-0.6 mg/ml and 1-2 mg protein/mg wet gel.

The unbound protein has been removed by repeated washing with 10 volumes of 100 mM potassium phosphate (pH 7.4) containing 0.9% NaCl.

To avoid inactivation of immobilized microsomes by excess of unreacted glutaraldehyde and by free radicals produced during the polymerization of acrylamide, the gels were incubated during 12 h at 25°C with 20 mM hydrazine and 100 μM hydroxybuthylated toluene, respectively, in phosphate buffer (pH 7.4). Then the gels were washed with phosphate buffer as indicated above, and stored at -20°C until use. Before use the gels have been centrifuged 5 min at 1600 x g and 4°C.

RESULTS AND DISCUSSION.

The fibers used to prepare the hollow fiber bioreactor have been Romicon PM10 polysulphone anisotropic fibers with an internal ultrafiltration membrane having a molecular weight cut-off of 10 kDa. Seven of these fibers were mounted on plastic pieces, fixed with an epoxy-type resine and pretreated as indicated by the manufacturer, e.g. 30 min incubation with 50 mM phosphoric acid, followed by 30 min washing with distilled water and 30 min incubation with 125 mM NaOH. Finally, the fibers were washed with distilled water and equilibrated with the buffer to be used (100 mM potassium phosphate/100 mM KCl/1 mM EDTA/ 1 mM β-mercaptoethanol and 50 μM sodium azide, pH 7.4).

Mounted fibers were installed in a close circuit (see the

Figure 1), and operated in transverse mode, i.e., medium from a reservoir containing buffer plus substrates passes into the shell side of the reactor and crosses through the walls of the fibers where it interacts with the immobilized cytochrome P450, and leave the reactor through the lumen of the fibers, then returning to the reservoir.

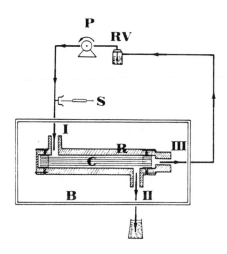

Figure 1. Scheme of the bioreactor (R) and the recycling system used in the present study. The medium (buffer and substrates) goes into the reactor through **port I**, crosses the wall of the fibers (installed in a cartridge configuration (C)), and leave the reactor through the lumen (**port III**). **Port II** is used to wash the shell side of the reactor. The immobilized system is immersed in a thermosthatic bath (**B**). The flow rate is selected and adjusted by means of a peristaltic pump (**P**). The activity is measured by taking and returning aliquots from the reservoir (**RV**) at different times. The protein to be immobilized is loaded in the reactor through the port marked as **S**. The following data refer to the average dimensions of the inner part of the bioreactor: total fibers mounted = 7; total fiber lenght = 84 cm; total fibers external surface = 53 cm^2.

The protein has been immobilized in the fibers by recirculating in transverse mode a diluted solution of microsomes at a flow rate of 30 ml/h during 90 min. The excess of protein has been washed by pumping buffer at high flow rate (1.5 ml/min) through the shell side during 45 min. On average, the results reported in this paper have been obtained with 6 mg protein immobilized in bioreactors of the size indicated in the legend of the Figure 1.

After the addition of the substrates to the reservoir (NADPH and 7-ethoxycoumarin as electron source and substrate to be metabolized, respectively) the activity has been determined from measurements of the intensity of the fluorescence emission of aliquots taken at different time intervals. The aliquots were then returned back to the reservoir. This sampling method was done until the fluorescence intensity reached a maximum and stable value. The level of NADPH has been maintained constant in the system by controlled additions of this nucleotide to the reservoir every 20-min.

Figure 2 illustrates a typical set of values obtained during the operation of the hollow fiber bioreactor.

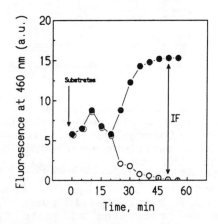

Figure 2. Sampling method used in the measurement of the activity of microsomal cytochrome P450 immobilized in hollow fiber bioreactors. The substrates (NADPH and 7-ethoxycoumarin) were added at the time indicated by the arrow. The net fluorescence increase (e.g. IF) has been used in the calculation of the product (7-hydroxycoumarin) formed by the enzyme (see the Methods). The different symbols correspond to: (●) bioreactors with cytochrome P450 and (○) bioreactors operated without protein.

At the time indicated by an arrow the substrates, 7-ethoxycoumarin and NADPH, were added. The transient increase of fluorescence observed immediatly afterwards is likely to be artifactual, probably related to the dispersion of the dimethylsulfoxide solution of 7-ethoxycoumarin in buffer. The aliquots were routinely taken every 2 or 5 min, and production of hydroxycoumarin was calculated from the net fluorescence increase at a given time, e.g. substracting the fluorescence reading of the experiment carried out in the absence of an electron donor compound.

From results like those presented in the Figure 2 we have calculated the deethylase activity of the immobilized microsomes from the initial slope of the raising phase. This activity, averaged from data obtained with 15 different bioreactors, amounts to 0.025 + 0.005 nmols product/min/mg protein at a flow rate of 0.5 ml/min and at saturation of 7-ethoxycoumarin. This activity is about 15-20% of the maximum activity of the native microsomes assayed in the same experimental conditions. We have obtained roughly the same value of $K_{0.5}$ 80-85 mM, for the substrate 7-ethoxycoumarin in immobilized and native microsomes (results not shown).

A very possitive feature of these bioreactors is the stability of the activity of immobilized cytochrome P450. These bioreactors retain more than 90% of the initial activity after 24 h at 25°C. The half-time of inactivation at 25°C of the cytochrome P450 in native microsomal membranes has been found to be aproximately 65 h. In addition, these bioreactors can be frozen and stored during at least one month without significant loss of enzyme activity.

Additionally, these hollow fiber bioreactors can be modified to measure the activity of the microsomal cytochrome P450 using an enzymatic system to generate electrons "in situ". This has been achieved coimmobilizing glucose-6-phosphate dehydrogenase with the microsomes. After washing the excess of protein and determining the amount of enzyme loaded as indicated above, the activity was measured in these conditions using NADP, glucose-6-phosphate and 7-ethoxycoumarin as substrates.

These latter two features are significant advantages with

respect to alternative methods of immobilization of this system.
To strenghten this point further we have immobilized microsomes
containing cytochrome P450 in acrylamide gels using
glutaraldehyde as the crosslinking agent. The pretreatment of
acrylamide (to form the solid support) is indicated in the
Methods. The extent of immobilization is summarized in the
Table I.

Table I. Effect of the concentration of acrylamide and
glutaraldehyde on the binding of hepatic microsomal cytochrome
P450 to acrylamide gels. The results shown corresponds to the
average of three immobilizations.

	Cyt c [3]	[ACRYLAMIDE] [1]		[GLUTARALDEHYDE] [2]	
		10%	15%	6%	12%
Incubated gel [4]	5	5	5	5	5
Incub. protein [5]	5	10	10	5	5
[Prot] incub. [6]	1	2	2	1	1
Bound prot. [5]	4.51	2.20	2.44	1.43	1.90
[Prot] bound [6]	1.29	0.44	0.49	0.29	0.38

(1) Glutaraldehyde concentration 6% (w/v).
(2) Acrylamide concentration 10% (w/v).
(3) Cytochrome c using concentrations of acrylamide and
glutaraldehyde of 10% and 6% (w/v), respectively.
(4) ml of wet gel after centrifugation at 1600 x g during 10 min
(5) total mg of protein.
(6) mg of protein/ml of wet gel after centrifugation at 1600 x g
during 10 min.

Figure 3 shows the effect of acrylamide and glutaraldehyde on
the dependence of the activity of the immobilized cytochrome
P450 upon the concentration of 7-ethoxycoumarin. From these
results a value of 80 ± 10 μM is obtained for the $K_{0.5}$ of this
substrate, which is identical within experimental errors to the
$K_{0.5}$ value of the microsomal cytochrome P450.

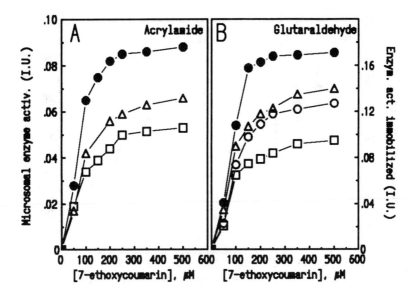

Figure 3. Dependence upon the concentration of 7-ethoxycoumarin of the activity of microsomal cytochrome P450 from phenobarbital treated rats immobilized by covalent coupling to acrylamide gels. Temperature: 25°C.
(1 I.U.: 1 nmol of 7-hydroxycoumarin produced/min/mg protein).
Panel A: Glutaraldehyde concentration used for covalent crosslinking 6% (w/v). The different symbols correspond to the following concentrations of acrylamide (w/v): (●) 10%; (△) 20% and (□) 30%.
Panel B: Acrylamide gels used 10% (w/v). The different symbols correspond to the following concentrations of glutaraldehyde (w/v): (○) 3%; (●) 6%; (△) 9% and (□) 12%.

Figure 4 shows the dependence of the enzymatic activity upon acrylamide and glutaraldehyde concentrations in the immobilization protocol. From the results shown in the Figure 4 it is evident that the immobilization in acrylamide gels leads to a maximum activity of the immobilized cytochrome P450 of ca. 15% of the maximum activity of the native microsomal system. Nevertheless, the microsomes immobilized in acrylamide gels show a half time of inactivation of aproximately 32 h at 20°C. This result indicates that this immobilization leads to a decreased stability of cytochrome P450, for the half time of this enzymatic activity at 25°C in native microsomal preparations is approximately 65 h.

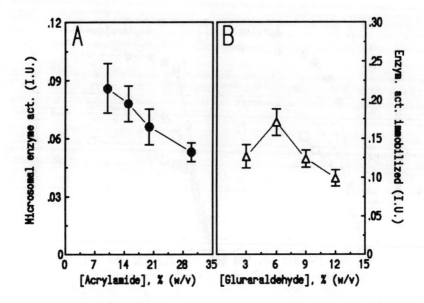

Figure 4: Effect of the concentration of acrylamide (Panel A) and glutaraldehyde (Panel B) on the maximum activity of microsomal cytochrome P450 from phenobarbital treated rats immobilized by covalent coupling to acrylamide gels. The data have been obtained from the experimental series shown in the Figure 3.
(1 I.U.: 1 nmol of 7-hydroxycoumarin produced/min/mg protein).

The electron microscopy pictures shown in the Figure 5 reveal the basic structural characteristics of the immobilized systems used in this study.

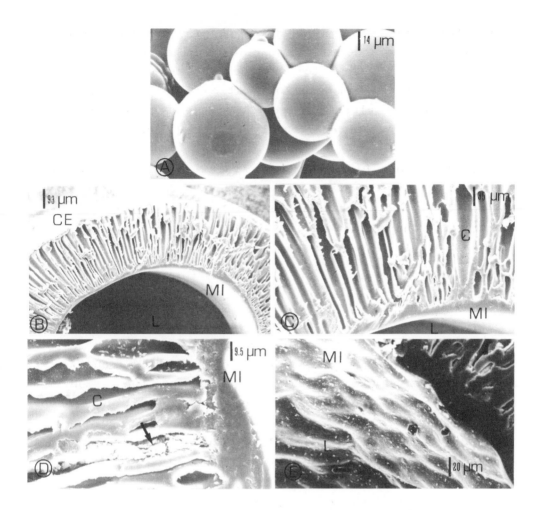

Figure 5: Scanning electron microphotographs of the supports
used in the immobilization of microsomal cytochrome P450.
(A) Acrylamide gels 10% (w/v); magnification: 500 x
(B-C-D-F) Hollow fiber. The abreviations correspond to: (MI)
inner membrane of the fiber; (C) channel in the membrane wall;
(CE) external surface of the fiber and (L) lumen of the fiber.
The cytochrome P450 microsomal aggregates are indicated by arrows
in D. The magnifications used have been: 75 x (B); 200 x (C);
750 x (D) and 350 x (F).

Regarding the acrylamide gels it is to be noted the homogeneous appearance of the particles. The sectional cut of the hollow fibers allows to envisage the mechanism of physical trapping on the microsomes. In addition, the distortions upon operation of the ultrafiltration membrane facing the luminal side of the bioreactor are readily evident (compare pictures 5F and 5B).

In hollow fiber bioreactors the value of the efectiveness factor of the reaction (η) yields a parameter related to the ratio between diffusional and kinetic control in these type of bioreactors (Chambers et al., 1976). This parameter η is defined by the equation,

$$v = (\eta \cdot V \cdot V_{max} \cdot S / (S + K_m))$$

where, v is the overall reaction rate; V, the total void volume of the bioreactor in the enzyme compartment (ml); S, the substrate concentration in the reaction medium; V_{max} and K_m, the maximum velocity and Michaelis constant of the immobilized enzyme. From the electron microscopy pictures, the parameter V has been estimated in our hollow fibers to be approximately $0.67 \times V_o$, being V_o the total volume occupied by the thick walls of the fibers. From titrations with 7-ethoxycoumarin of the activity of native microsomes and of microsomes immobilized in hollow fiber (not shown) we have obtained the values of K_m, V_{max} and v at several substrates concentrations (S). From these results we have calculated a value of η of 0.9-1.0, and, thus, we conclude that the bioreactor operates mostly under kinetic control at a flow rate of 0.5 ml/min. This conclusion is consistent with the fact that the value of $K_{0.5}$ for 7-ethoxycoumarin of the immobilized microsomes is identical to that of native microsomes (see above).

The effect of the pesticides warfarin and amino-triazol on the activity of the cytochrome P450 immobilized in hollow fiber is illustrated in the Figure 6.

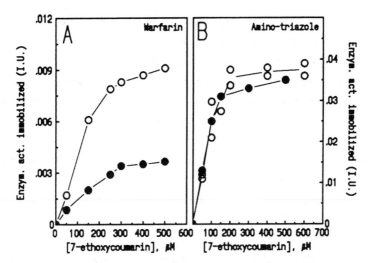

Figure 6: Effect of warfarin and amino-triazole on the dependence upon the concentration of 7-ethoxycoumarin of the activity of microsomal cytochrome P450 from phenobarbital treated rats immobilized in hollow fiber bioreactors. Temperature: 25°C. Flow rate:0.5 ml/min.
(1 I.U.: 1 nmol of 7-hydroxycoumarin produced x flow rate (ml/min)/mg protein).
Panel A: Warfarin.The different symbols correspond to the following concentrations of warfarin (μM): (○) 0 and (●) 120. Immobilized protein: 11.82 mg.
Panel B: Amino-triazole. The different symbols correspond to the following concentrations of amino-triazole (μM): (○) 0 and (●) 120. Immobilized protein: 5.55 mg.

It can be observed that concentrations of these pesticides close to their respective LD_{50} doses differ in their actions on the cytochrome P450, for amino-triazol does not have any significant effect on this system and warfarin behaves as a non-competitive inhibitor. This latter point was assessed further using different concentrations of warfarin, and confirmed with native microsomes (results not shown).

CONCLUSIONS AND PERSPECTIVES. The results presented in this paper shows that the cytochrome P450 system of hepatic microsomes can be immobilized in hollow fiber bioreactors. The immobilized system shows a maximum activity of about 15-20% of that in native membranes and an enhanced stability, without

304

significant loss of activity during 24 h of continuous operation
and upon storage for several weeks frozen at -20°C. In addition
the catalytic efficiency of the cytochrome P450 can be slightly
improved by the coimmobilization with systems capable to
generate electrons "in situ", such as glucose-6-phosphate
dehydrogenase. As discussed above,these are practical advantages
of hollow fibers bioreactors over other immobilization
approaches, such as poliacrylamide beads. We have also shown
that the hollow fiber bioreactors can be used as a model system
to study the effects of pesticides upon the catalytic activity
of the cytochrome P450 system.

Because hollow fiber bioreactors are successfully used to get
urea clearance from the blood (hemodialysis), it is tempting to
speculate that hollow fiber bioreactors of cytochrome P450 can
be used for drugs and pesticides detoxification of the blood of
human beings. In this regard, it is to be noted that the
adaptative induction of isoenzymes of cytochrome P450 in
response to chemical insult by different organisms make
a realistic possibility to produce a large number of bioreactors
with varying catalytic efficiency to different substrates.

Acknowledgements: This work has been supported, in part, by a
Grant of the Junta de Extremadura, and by a fellowship of the
Spanish Ministerio de Educación y Ciencia (Biotechnology
Program) to Dr. P. Fernandez-Salguero.

REFERENCES

Åström, A., Birberg, W., Pilotti, Å., and DePierre, J.W. (1985)
Eur. J. Biochem. 154, 125-134
Bunch, A.W. (1988) J. Microb. Methods 8, 1103-119
Chambers, R.P., Cohen, W., and Baricos, W.H. (1976), In: Methods
in Enzymology, vol. XLIV (Mosbach, K., ed), Academic Press, New
York, pp. 291-317
Coulson, C.J., King, D.J., and Wiseman, A. (1984) Trends in
Biochemical Sciences 9, 446-449
Estrada-Diaz, P., Sallis, P.J., Bunch, A.W., Bull, A.T., and
Hardman, D.J. (1989) Enz. Microb. Technol. 11, 725-729
Lowry, O.H., Rosebrough, H.J., Farr, A.L., and Randall, R.L.
(1951) J. Biol. Chem. 193, 265-275
Lu, A.Y.H., and Levin, W. (1972) Biochem. Biophys. Res. Commun.
46, 1334-1339

Nebert, D.W., and Gonzalez, F.J. (1987) Ann. Rev. Biochem. 56, 945-993

Omura, T., and Sato, R. (1964) J. Biol. Chem. 239, 2370-2378

Pompon, D. (1987) Biochemistry 26, 6429-6435

Tredger, J.M., Smith, H.M., Davis, M., and Williams, R.T. (1984) Biochem. Pharmacol. 33, 1729-1737

Waxman, D.J., and Walsh, C. (1982) J. Biol. Chem. 257, 10446-10457

White, R.E., Oprian, D.P., and Coon, M.J. (1980), In: Microsomes, Drug Oxidations and Chemical Carcinogenesis, vol. I (Coon, M.J., Conney, A.H., Estabrook, R.W., Gelboin, H.V., Gillette, J.R., and O'Brien, P.J., eds), Academic Press, New York, pp. 243-251

Progress in Membrane Biotechnology
Gomez-Fernandez/Chapman/Packer (eds.)
© 1991 Birkhäuser Verlag Basel/Switzerland

THE BINDING PARAMETERS OF A HYDROPHOBIC CATIONIC DRUG TO PHOSPHOLIPID MEMBRANES

Peter G. Thomas♥ & Joachim Seelig♦

♥ Institute of Molecular Biology and Medical Biotechnology, University of Utrecht, Utrecht,The Netherlands

♦ Dept. of Biophysical Chemistry, Biozentrum, University of Basel, Basel, Switzerland

SUMMMARY:Flunarizine binding to phosphatidylcholine, under conditions where the drug possesses a positive charge, displays non-linear characteristics. This non-linear binding can be explained electrostatically. Application of the Gouy-Chapman theory permits the calculation of the true interfacial aqueous concentration of flunarizine which gives rise to linear concentration dependent binding. This allows the derivation of a coefficient for the binding of flunarizine to phosphatidylcholine of 26020 M^{-1}. This is a very large value and supports the notion that this drug may be particularly membrane-active.

INTRODUCTION

The interaction of cationic amphiphilic drugs with membranes has long been of interest to the membrane biochemist. The pharmacological activity of such molecules usually takes place at the plasma membrane level and frequently involves an initial interaction of the drug with the membrane phospholipids even if those phospholipids are not the final 'target' of the drug molecule.

It is generally thought that these drugs, when charged, are intercalated into the phospholipid bilayer with their hydrophobic moieties inserted into the acyl chain region and the charged groups positioned around the glycerol backbone/phosphate region, i.e. the 'membrane interface'. This has been shown to be the case for quite a number of pharmacologically unrelated amphiphilic drugs, including tetracaine (Boulanger et al., 1981), propranolol (Herbette et al., 1983), chlorpromazine (Kuroda & Kitamura, 1984), and nimodipine (Herbette et al., 1986). By insertion into the phospholipid bilayer these drugs are thereby able to exert a significant influence upon the phase behaviour of phospholipids. Differential scanning calorimetry (DSC) has been used to assess the influence of a variety of drugs on the phase behaviour of in a wide range of studies (see references 11-17 in Thomas & Verkleij, 1990). We have recently used DSC to investigate the influence of one such cationic amphiphilic drug, flunarizine, on the phase behaviour of different phospholipid classes and molecular species (Thomas & Verkleij, 1990).

Flunarizine (Fig 1) is a class IV calcium antagonist; that is to say that unlike the other three classes of calcium antagonist it has no proven specific binding with the voltage gated calcium slow channel. Flunarizine appears to be able to block 'calcium overload' (see Todd & Benfield, 1989) thus preventing cellular damage under stress conditions. Indeed, we have shown that flunarizine is able to prevent damaging

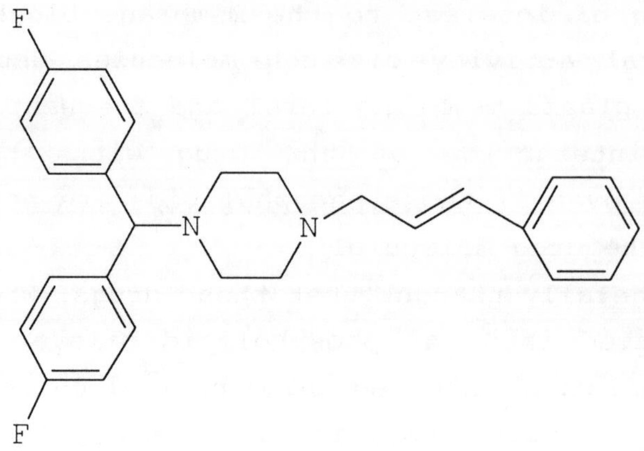

Fig 1: Structure of flunarizine (1-[*bis*(4-fluorophenyl)methyl]-4-(3-phenyl-2-propenyl)piperazine).

calcium-induced membrane reorganization from occurring under conditions of high calcium concentration (Thomas et al., 1988).

Owing to the greatly differing chemical structures of this class of molecule it is inevitable that they have quite different physicochemical properties. Unlike many calcium antagonists flunarizine is very hydrophobic, even in its charged state. In order to better understand the interaction of flunarizine with phospholipids we have investigated the binding of this drug to phosphatidylcholine (PtdCho) bilayers using centrifugation assays and zeta potential measurements. These binding studies demonstrate the complex nature of the interaction of a charged amphiphilic molecule with phospholipid bilayers.

Recently the Gouy-Chapman theory has been used to examine the binding of charged molecules to phospholipid bilayers (Seelig et al., 1988, Bäuerle & Seelig, 1991). We have applied this theory to examine the binding of flunarizine to PtdCho bilayers. The results reveal a very high binding coefficient with binding characteristics that can be explained electrostatically by the Gouy–Chapman theory.

MATERIALS AND METHODS

The experiments were carried out using 1–palmitoyl–2–oleoyl–sn–glycero–3–phosphocholine (Pam$_1$Ole$_2$PtdCho) which was purchased from Avanti Polar Lipids (U.S.A.) in dry form. Flunarizine was a gift from Janssen Pharmaceutica (Belgium).

Sample preparation: Lipid–drug dispersions were prepared as follows. Briefly 4 mg Pam$_1$Ole$_2$PtdCho were suspended in 5 cm^3 of a series of different concentrations of flunarizine (50 – 100 μM) in 100 mM NaCl, 30 mM sodium citrate/hydrogen phosphate buffer (pH 5.0). Experiments were carried out at pH 5.0 to ensure that flunarizine (pK$_a$ = 7.7, Janssen Pharmaceutica, Analytical Department, Beerse, Belgium) was in its charged form.

    Drug–lipid dispersions were also prepared by freeze–thaw and extrusion methods. No differences in drug binding to PtdCho were found using these different methods. The results presented here were all obtained from simple dispersion of the phospholipid in drug containing buffer.

ζ–potential measurements: Large multilamellar lipid dispersions were prepared as described above. ζ–potential measurements were carried out at 25°C using a Rank Brothers Mark II microelectrophoresis apparatus (Cambridge, U.K.). Observations were made at the stationary level of the cell where there is effectively no net solvent flow. The average of 16 measurments in both directions was used to estimate the electrophoretic mobility, u. The ζ-potential was calculated from the Helmholz–Smoluchowski equation:

$$\zeta = \eta u / \varepsilon_r \varepsilon_0$$

where η is the viscosity, ε_r the dialectric constant and ε_0 the permittivity of free space (see McLaughlin, 1977). Drug concentration at equilibrium was estimated as described below.

Centrifugation binding assays: Large multilamellar dispersions were prepared as described above. The drug–lipid dispersions were centrifuged at 20°C for 2 h at 340,000g giving a clear supernatant. After careful removal of the supernatant the equilibrium concentration of flunarizine in solution (c_{eq}) was determined spectrophotometrically ($\varepsilon_{253}=21900$ $M^{-1}cm^{-1}$, Janssen Pharmaceutica, Analytical Department, Beerse B). The amount of bound flunarizine could then be calculated by subtraction of c_{eq} from the starting concentration of the drug. The mole fraction of bound drug, X_b, is defind as

$$X_b = n_D / n_L$$

where n_D drug molecules are bound to a total of n_L lipid molecules at a given drug concentration.

RESULTS

Incorporation of flunarizine into PtdCho bilayers under conditions of low pH where it possesses a positive charge (electrical charge, $z = +1$) causes a charging up of the phospholipid bilayer. This can be measured electrophoretically by ζ-potential determination demonstrating that the phospholipid bilayers become positively charged as the drug is taken up from the aqueous phase (Fig 2).

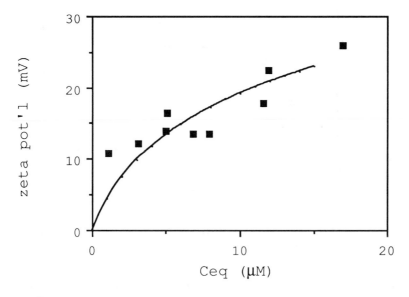

Fig 2: ζ-potential measurements of PtdCho multilamellar vesicles at various concentrations of flunarizine. The theoretical curve was calculated using $K_p = 26020$ M^{-1} and $z = +1$ taking electrostatic effects into account by means of the Gouy-Chapman theory (see Discussion).

The results of centrifugation binding assays are shown in Fig 3. They demonstrate non-linear characteristics for flunarizine binding indicating that less drug is taken up at higher concentrations than might be expected. This deviation away from linear concentration-dependent binding can be explained by consideration of the influence of the drug's positive charge. As more and more charged drug molecules are taken up by the bilayer the bilayer itself will become positively charged. This charge will repel the like-charged drug molecules in the aqueous phase close to the phospholipid bilayer leading to a concentration gradient of the drug with

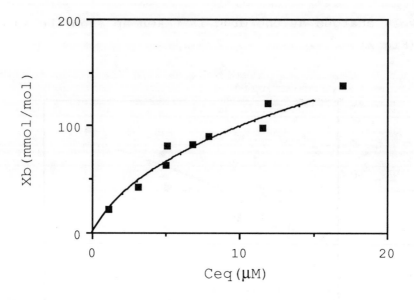

Fig 3: The binding of flunarizine (z = +1) to PtdCho bilayers. The extent of binding (mmol flunarizine per mol phospholipid) is plotted against the equilibrium concentration of flunarizine in the bulk solution. The solid line represents the predicted binding curve calculated using the Gouy–Chapman theory with K_p = 26020 M^{-1} and z = +1 (see Discussion).

the surface concentration of the drug being lower than that some distance away (≥2 nm) in the bulk solution. Therefore, particularly at higher concentrations, the more that the drug is taken up the greater the effect of charge repulsion on the surface concentration of the drug. The Gouy–Chapman theory can be used to describe such an exponential reduction in the concentration of ions in free solution approaching a charged surface of like charge (see McLaughlin, 1977). Application of the Gouy–Chapman theory allows the calculation of the true concentration of charged solutes immediately above a charged surface via the determination of the surface potential, Ψ_0.

$$\sigma^2 = 2000\epsilon_o\epsilon_r RT \Sigma c_{i,eq}(e^{\frac{-z_i F_o \Psi_o}{RT}} - 1)$$

where σ is the surface charge density, $c_{i,eq}$ is the concentration of the i-th electrolyte in the bulk aqueous phase, z_i is the signed valency of the i-th species and F_o the Farraday constant.

$$C_m = C_{eq}\exp\left(\frac{-F_o\Psi_o}{RT}\right)$$

Using this concentration (interfacial concentration at the membrane surface, c_m) instead of the equilibrium concentration in bulk solution generates a linear relationship between extent of drug binding and drug concentration in solution; $X_b = K_p c_m$ (Fig 4). The 'true' partition coefficient can then be easily calculated and is found to be K_p = 26020 M^{-1}.

DISCUSION

In this study we have examined the binding of a cationic amphiphilic drug, flunarizine, to phospholipid bilayers. The results presented here clearly demonstrate that such binding is more complex than might be expected. When the positively charged flunarizine binds to the PtdCho bilayers they themselves become charged as is demonstrated by ζ-potential measurements. This 'charging-up' of the bilayers has a large influence on the binding characteristics causing a deviation from linear concentration-dependent binding because of

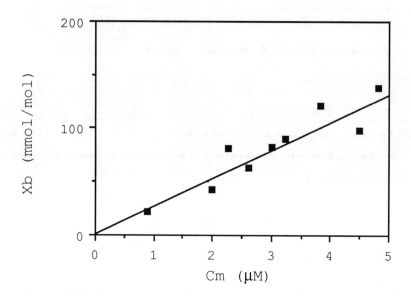

Fig 4: Flunarizine binding to PtdCho bilayers. The extent of flunarizine binding (mmol flunarizine per mol phospholipid) is plotted against the interfacial concentration of flunarizine. The solid line corresponds to $X_b = K_p c_m$ where $K_p = 26020$ M^{-1}.

electrostatic repulsion as demonstrated in Fig 3.

Application of the Gouy-Chapman theory allows the derivation of the surface charge density and the surface potential. From these values the influence of surface charge on the interfacial concentration of charged species can be calculated. It is thereby possible to determine the true concentration of flunarizine at the bilayer-water interface. A plot of these values of flunarizine concentration reveals linear concentration-dependent binding (Fig 4) and allows the derivation of the binding constant.

By using this binding constant it is possible to generate synthetic data for ζ-potentials and X_b at specified free drug concentrations. Figures 2 and 3 show that there is excellent agreement betweeen the theoretically derived data (solid lines) and that derived experimentally. We can see that the Gouy-Chapman theory, despite criticism for its oversimplification of the true situation (see Cevc, 1990), does allow good approximation of the binding of this positively charged drug to PtdCho bilayers.

The derived binding constant of $K_p = 26020$ M^{-1} is a very large value and is indicative of the strong interaction of flunarizine with phospholipids. Indeed, more than 95% of the drug has partitioned into the bilayers at the lower concentrations tested. These results support the notion that flunarizine is a particularly membrane-active drug. The mechanism of action of flunarizine remains a mystery at this moment, however, the results presented here give credibility to the idea that flunarizine may exert its action via its influence on cell membrane constituents, be it the membrane phospholipids or other membrane components.

Conclusions: These results clearly demonstrate that electrostatic interactions have a considerable effect on the binding of charged molecules to phospholipid bilayers. At first sight it would not seem to be possible to easily generate a meaningful binding constant from such data. However, application of the Gouy-Chapman theory allows the calculation of the true concentration of free drug at the bilayer surface. It can be seen, by using these drug concentration values, that flunarizine binding does display simple linear binding characteristics once unmasked of electrostatic influences.

Acknowledgements: P.G.T. gratefully acknowledges the financial support of The Janssen Research Foundation (Neuss, Germany).

REFERENCES

Bäuerle, H.-D. & Seelig, J. (1991) Biochemistry *in the press*.

Boulanger, Y., Schreier, S. and Smith I.C.P. (1981) Biochemistry 20, 6824–6830.

Cevc, G. (1990) Biochim. Biophys. Acta 1031, 311–382.

Herbette, L.G., Katz, A.M. and Sturtevant, J.M. (1983) Mol. Pharmacol. 24, 259–269.

Herbette, L.G., Chester, D.W. and Rhodes, D.G. (1986) Biophys. J. 49, 91–93.

Kuroda, Y. and Kitamura, K. (1984) J. Am. Chem. Soc. 106, 1-6.

McLaughlin, S. (1977) Curr. Topics Membrane Transport 9, 71–144.

Seelig, A., Allegrini, P.R. and Seelig, J. (1988) Biochim. Biophys. Acta 939, 267–276.

Thomas, P.G., Zimmermann, A.G. and Verkleij, A.J. (1988) Biochim. Biophys Acta 946, 439–444.

Thomas, P.G. and Verkleij, A.J. (1990) Biochim. Biophys. Acta 1030, 211–222.

Todd, P.A. and Benfield, P. (1989) Drugs 38, 481–499.

Progress in Membrane Biotechnology
Gomez-Fernandez/Chapman/Packer (eds.)
© 1991 Birkhäuser Verlag Basel/Switzerland

MEMBRANE FUSION ACTIVITY OF INFLUENZA VIRUS AND RECONSTITUTED VIRAL ENVELOPES (VIROSOMES)

Jan Wilschut[1], Romke Bron[1], Jan Dijkstra[1], Antonio Ortiz[1§], Lucina C. van Ginkel[1¶], William F. DeGrado[2], Maria Rafalski[2] and James D. Lear[2]

[1]Laboratory of Physiological Chemistry, University of Groningen, Bloemsingel 10, 9712 KZ Groningen, The Netherlands.
[2]The DuPont-Merck Pharmaceutical Co., P.O. Box 80328, Wilmington, DE 19880-0328, USA.
[§]Present address: Department of Biochemistry and Molecular Biology, University of Murcia, E-30071 Murcia, Spain.
[¶]Present address: Department of Plant Physiology, University of Groningen, Kerklaan 30, 9751 NN Haren, The Netherlands.

SUMMARY: Influenza virus enters its host cell through receptor-mediated endocytosis, routing the virus particles into the endosomal cell compartment. Induced by the mildly acidic pH within the endosomes, the viral membrane fuses with the endosomal membrane, resulting in release of the nucleocapsid in the cellular cytoplasm. This fusion process is mediated by the viral spike glycoprotein hemagglutinin (HA), which undergoes a series of conformational changes at low pH, exposing the hydrophobic N-terminus of the HA2 subunit. Synthetic peptides corresponding to this "fusion peptide" destabilize model membranes by penetration into the lipid bilayer in an α-helical conformation. Reconstituted influenza virus envelopes (virosomes), prepared by a $C_{12}E_8$ detergent solubilization/removal procedure, exhibit fusion characteristics very similar to those of the native virus. These virosomes are a useful model system for the intact virus in studies on viral fusion and cellular entry. In addition, the virosomes provide a promising carrier system for the introduction of biologically active, foreign molecules into living cells.

INTRODUCTION

Influenza virus is an enveloped virus, i.e., the viral nucleocapsid, containing the genetic material, is surrounded by a
lipid bilayer membrane. The virions acquire this membrane as
they bud from the plasma membrane of an infected host cell.
Initiating a subsequent round of cellular infection, enveloped
viruses utilize a membrane fusion strategy to deposit their
genome into the cytoplasm of new host cells (for reviews, see
Marsh and Helenius, 1989; Stegmann et al., 1989b; Wilschut and
Hoekstra, 1990; White, 1990). This fusion reaction may either
occur at the level of the host cell plasma membrane, or from
within endosomes after uptake of intact virions through
receptor-mediated endocytosis (Figure 1). In the latter,
endocytic, route of entry, the target membrane for the fusion of
the viral envelope is the limiting membrane of the endosomal
cell compartment. In a subcategory of enveloped viruses,
membrane fusion capacity is activated only under conditions of a
mildly acidic pH. These low-pH-dependent viruses must utilize
the endocytic route of cellular infection: Fusion at the plasma
membrane is precluded by the strict pH dependence of their
fusion activity, while, conversely, after endocytic uptake, the
virions encounter an acidic environment in the lumen of the
endosomes (Mellman et al., 1986). Infection of cells by these
viruses can be blocked by inhibitors of vacuolar acidification,
such chloroquin or NH_4Cl. In model systems, low-pH-dependent
viruses can be made to fuse with a variety of target membranes
by a transient lowering of the pH in the medium. Under these
conditions low-pH-dependent viruses can even be induced to fuse
with the plasma membrane of cultured cells, and, although this
fusion process is not "physiological", it occasionally leads to
infection of the cell. Viruses for which the expression of their
membrane fusion activity is not dependent on low pH, such as
e.g. HIV (McClure et al., 1988), in principle have the ability
to fuse at the plasma membrane under physiological conditions.
Whether or not these viruses also utilize the endocytic route of
infection is not known in many cases, including that of HIV.

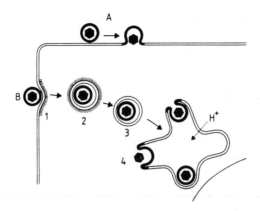

<u>Figure 1</u>:   Entry of enveloped viruses into cells.   A: Fusion of
the viral envelope with the plasma membrane; pH-independent
viruses may utilize this route of infection, whereas low-pH-
dependent viruses can only be made to fuse with the plasma
membrane of cultured cells by lowering of the extracellular
pH. B: Uptake of intact virions by receptor-mediated endocytosis through coated pits (1), routing the virions into
coated vesicles (2), which after uncoating (3), fuse with
endosomes, where, in the case of low-pH-dependent viruses,
the acidic pH activates the fusion activity of the virus (4).
(Reproduced, with permission, from Wilschut et al., 1988).

It is well established that influenza virus utilizes the
endocytic route to infect its host cell (for reviews, see
Stegmann et al., 1989b; White, 1990; Doms et al., 1990). The
membrane of the virus contains two major integral spike
glycoproteins, the hemagglutinin (HA) and the neuraminidase
(NA); in addition, there is the minor integral membrane protein
M2. The infectious entry of the virions into the host cell is
mediated by the HA.   First, HA binds to sialic-acid-containing
receptors on the cell surface. Second, following the internali-
zation of the virus particles into the endosomal cell compartment (Matlin et al., 1981; Stegmann et al., 1987a), the HA also
triggers the fusion reaction with the endosomal membrane.

HA is the best characterized membrane fusion protein so far
(for reviews, see Wiley and Skehel, 1987; Stegmann et al.,
1989b; Doms et al., 1990; White, 1990). The HA spike, protruding
some 13.5 nm from the viral surface, is a homotrimeric molecule.
Each monomer consists of two disulfide-linked subunits, HA1 (47
kD) and HA2 (28 kD), which are generated from a single poly-

peptide chain, HA0 (85 kD), by posttranslational cleavage
involving a host-cell protease. The globular HA1 domains contain
the sialic-acid binding pockets. The N-terminus of HA2 appears
crucial for the expression of fusion activity of HA: Uncleaved
HA0 is not fusion-active, while site-specific mutations within
this region of the molecule severely affect the fusion activity
of HA (Gething et al., 1986). The N-terminus of HA2, the so-
called "fusion peptide", is a conserved stretch of some 20 amino
acid residues, that are mostly hydrophobic in nature (White,
1990). At neutral pH the fusion peptides are buried within the
stem of the HA trimer at about 3.5 nm from the viral surface.
However, at low pH an irreversible conformational change in the
HA (Skehel et al., 1982; Doms et al., 1985) results in their
exposure (White and Wilson, 1987). Interaction of the fusion
peptides with the target membrane then probably triggers the
fusion event.

In this review we will first briefly discuss assays for
monitoring viral fusion in model and cellular systems, including
an introduction to the use of reconstituted viral envelopes
(virosomes). Then several examples of influenza virus and
virosome fusion are presented, followed by a discussion of the
possible molecular mechanism of influenza virus fusion. Finally,
we address the issue of application of virosomes as a carrier
system for the introduction of foreign molecules into cells.

FLUORESCENCE ASSAYS FOR MONITORING VIRAL FUSION

Fusion of enveloped viruses can be monitored in a variety of
model systems using fluorescence assays that permit a kinetic
assessment of the process.

RET Assay: In model systems, involving the fusion of intact
virions with artificial lipid vesicles (liposomes), a
fluorescence assay relying on resonance energy transfer (RET)
between N-NBD-PE and N-Rh-PE (Struck et al., 1981), has been
applied extensively. Upon fusion, the two fluorophores, pre-

incorporated in the liposomal membrane, dilute into the viral membrane, resulting in a decrease of their overall surface density and a concomitant decrease of the RET efficiency. This decrease is monitored continuously as an increase of the donor (N-NBD-PE) fluorescence. The RET assay is not suited for the assessment of the fusion of intact virus with biological membranes, since the fluorophores can not be incorporated into either the viral or the biological target membrane.

R_{18} Assay: Fusion of intact virus with a biological target membrane can be followed with an alternative assay (Hoekstra et al., 1984), relying on the self-quenching properties of octadecyl Rhodamine B (R_{18}). Intact virions are labeled to a self-quenching surface density of R_{18}; fusion is monitored as an increase of the fluorescence due to relief of this self-quenching as the probe dilutes into the target membrane.

Even though the R_{18} assay, in a number of model systems, has proven to reliably reveal the general fusion characteristics of several enveloped viruses, there is reason for some concern regarding the assay. This concern is based, in part, on the following (unpublished) observations: (i) the probe, after labeling of the virions, appears to be not uniformly distributed in the viral membrane, and (ii) the rate and extent of fusion of reconstituted virosomes (see below) measured with the R_{18} assay are lower than the same parameters measured with an alternative assay for lipid mixing, the pyrene-PC assay (see below). This suggests that, after fusion, the R_{18} probe may dilute incompletely and/or in a retarded fashion into the target membrane.

Pyrene-PC Assay: This assay relies on the capacity of the pyrene fluorophore to form excimers between a probe molecule in the excited state and a probe molecule in the ground state. The fluorescence emission of the excimer is shifted to higher wavelengths by about 100 nm relative to the emission of the monomer. Excimer formation is dependent on the distance between the probe molecules. Thus, coupled to one of the acyl chains of a phospholipid molecule, such as phosphatidylcholine (PC), the pyrene probe provides a sensitive measure of the surface density of the labeled molecule in a lipid bilayer membrane.

We have used pyrene-PC to follow the interaction of virosomes (see below) with model membranes and cells. Fusion is observed as a decrease of the excimer fluorescence. In principle, it is also possible to insert the probe, from liposomes as donor vesicles, into the membrane of intact virus particles by using a PC-exchange protein. Alternatively, virus can be produced from cultured cells, the membrane lipids of which being prior labeled biosynthetically with pyrene fatty acids (Pal et al., 1988).

Reconstituted Virosomes as a Model for the Native Virus: In order to elucidate the specific roles of the various viral membrane components in cellular entry and fusion, it is essential to isolate these components and to reconstitute them in a functional manner. A first step toward this goal has been the development of a procedure for the functional reconstitution of influenza spike glycoproteins in the viral lipids (Stegmann et al., 1987c; Wilschut and Stegmann, 1988). The viral envelope is dissolved in an excess of the non-ionic detergent $C_{12}E_8$, and after sedimentation of the nucleocapsid by ultracentrifugation, the detergent is removed from the supernatant by a two-step treatment with BioBeads SM2. The membrane fusion characteristics of the virosomes, thus produced, are very similar to those of the native virus. The virosomes can, therefore, serve as a model for the virus in studies on viral fusion and cellular entry, a model allowing the introduction of reporter molecules in the virosomal membrane or in the enclosed volume. Virosomes also provide a promising carrier system for introduction of foreign molecules into living cells.

FUSION OF VIRUS AND VIROSOMES WITH MODEL MEMBRANES AND CELLS

The membrane fusion activity of influenza virus has been characterized in detail in a various model and cellular systems, by application of the fluorescence assays, described above.

<u>Fusion with Liposomes</u>: Using the RET assay, Stegmann et al. (1985, 1987b, 1989a, 1990) have characterized extensively the fusion of influenza virus with artificial lipid vesicles. The virus fuses quite readily with liposomes. Fusion exhibits the characteristic pH dependence of the conformational change in the HA molecule (see below), provided that the target liposomes do not contain high concentrations of negatively charged phospholipids, such as e.g. phosphatidylserine (PS) or, particularly, cardiolipin (CL). Liposomes composed of these latter lipids appear to support a very efficient and fast fusion reaction with a variety of biological membranes, including viral membranes. Although the rate of this fusion process does increase with decreasing pH, this pH dependence is relatively shallow, resulting in significant rates of fusion of influenza virus with CL liposomes at neutral pH, i.e. above the threshold pH for the occurrence of the conformational change in the HA (Stegmann et al., 1985, 1986). These properties of influenza virus fusion with CL liposomes have led us to conclude that it is of a non-physiological nature (Wilschut et al., 1988; Stegmann et al., 1989a). On the other hand, fusion with zwitterionic target liposomes reveals the characteristics of the physiological process.

The presence of specific sialic-acid-containing receptors in the target membrane is not essential for fusion of influenza virus with zwitterionic liposomes, although the binding of the virus to the liposomes is enhanced when the liposomes contain, e.g., gangliosides (Stegmann et al., 1989a). Recent evidence suggests that gangliosides may also accelerate the formation of an active fusion complex (see below) and enhance the rate of the actual fusion process (Stegmann et al., 1990).

<u>Fusion with Erythrocyte Ghosts</u>: Figure 2 gives an example of influenza virus fusion with human erythrocyte ghosts. Both fusion of R_{18}-labeled intact virus, and fusion of pyrene-PC-labeled virosomes are shown. Although, in either case, fusion is strictly dependent on low pH, clearly, at the optimum pH for fusion (pH 5.5), the fusion of the virosomes appears faster and more extensive than that of the intact virus. However, virosomes

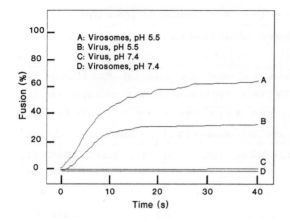

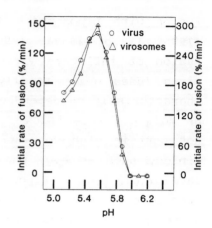

Figure 2 (left): Fusion of R_{18}-labeled influenza virus (X-99) and pyrene-PC-labeled virosomes with human erythrocyte ghosts. Virions/virosomes were allowed to bind to the ghosts for 15 min at 4 °C and pH 7.4; after washing, an aliquot of a concentrated suspension of the complex was then injected into medium of the desired pH at 37 °C in the cuvette of the fluorimeter. Virus fusion was monitored as an increase of the R_{18} fluorescence, virosome fusion as a decrease of the pyrene-PC excimer fluorescence.

Figure 3 (right): Initial rates of fusion of influenza virus and virosomes with erythrocyte ghosts. Conditions as in Figure 2.

labeled with the R_{18} probe, as compared to pyrene-PC-labeled virosomes, also exhibit a slower and less extensive fusion reaction with erythrocyte ghosts (not shown), indicating that the difference in fusion rates between virus and virosomes shown in Figure 2 is due to the use of the R_{18} assay in the case of the virus, rather than a reflection of a real difference in fusion potential between virus and virosomes. From extensive studies we conclude that the fusion characteristics of virosomes are very similar to those of the native virus. As an illustration, Figure 3 shows the detailed pH dependence for the two processes.

Fusion with Cultured Cells: Figures 4 and 5 present recent results on the fusion of influenza virus and virosomes with cultured BHK-21 cells. Unlike the approach followed in previous work on the interaction of the virus with cells (Stegmann et al., 1987a; Wilschut and Stegmann, 1988), here the fusion process was monitored on-line in the cuvette of the fluorimeter

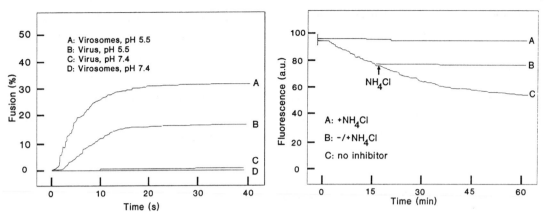

<u>Figure 4 (left)</u>: Fusion of R_{18}-labeled influenza virus (X-99) and pyrene-PC-labeled virosomes with the plasma membrane of BHK-21 cells in suspension. Virions/virosomes were allowed to bind to the cells for 1 h at 4 °C and pH 7.4; after washing, an aliquot of the cell suspension was injected into medium at 37 °C and the desired pH in the cuvette of the fluorimeter. Fusion was monitored as described in the legend to Figure 2.

<u>Figure 5 (right)</u>: Intracellular fusion of influenza virosomes with BHK-21 cells. The extracellular pH was 7.4. Otherwise, conditions were as in Figure 4. NH_4Cl conc., 20 mM (pH 7.4).

by using cells detached from the culture dish by a brief trypsinization. Figure 4 shows fusion at the cellular plasma membrane, induced by lowering of the extracellular pH, while Figure 5 presents intracellular fusion with the endosomal membrane at neutral extracellular pH. Clearly, fusion at the plasma membrane is a very fast process, starting after a lag of only a few seconds. Again, the R_{18}-labeled intact virus appears to fuse less rapidly and less extensively than the pyrene-PC-labeled virosomes (see also Figure 2). At neutral pH, at the time scale of the experiment, there is no change in fluorescence intensity (Figure 4). Under these conditions, the virions/virosomes are endocytosed and directed to endosomes; however, it takes several minutes before the particles reach this compartment from where they can fuse. This lag period is clearly indicated in Figure 5 (curve B/C). Depending on the pH threshold for fusion of the particular strain of virus used, this lag may vary from about 5 min (Figure 5) to 15 min (not shown). This is consistent with fusion from early and late endosomes, resp., the

latter having a lower lumenal pH than the former. The intracellular fusion reaction is completely blocked by NH₄Cl (Figure
5), an inhibitor of endosomal acidification (Mellman et al.,
1986). When added during the course of the fusion reaction,
NH₄Cl arrests the process almost instantaneously (Figure 5).
Conversely, when the NH₄Cl is washed away, the endosome lumen is
rapidly reacidified, upon which fusion resumes (not shown).

MOLECULAR MECHANISM OF INFLUENZA VIRUS FUSION

Conformational Changes in HA induced by Low pH: It is well
established that the expression of fusion activity of the HA
molecule is triggered by a series of low-pH-dependent conformational changes (Skehel et al., 1982, Doms et al., 1985; White
and Wilson, 1987; Stegmann et al., 1987b, 1989b, 1990). The
general features of these conformational changes are well
understood. The overall secondary structure of the molecule
remains largely unchanged. Also, the trimeric nature of the
spike is maintained. On the other hand, hidden antibody epitopes
and sites for proteolytic cleavage become exposed. Importantly,
the previously buried fusion peptides are exposed. Also, in the
final low-pH conformation the top of the spike is opened up due
to dissociation of the globular HA1 domains from each other.

During the course of the conformational change the bromelain-
solubilized ectodomain of HA (BHA) becomes amphiphilic (Doms et
al., 1985). It has been shown by hydrophobic photolabeling that,
at low pH, BHA interacts with liposomes through penetration of
the fusion segment into the liposomal bilayer (Harter et al.,
1988, 1989; Brunner, 1989). In addition, these latter studies
have indicated that the fusion peptide adopts an α-helical
structure during this interaction (Harter et al., 1989), and,
furthermore, that the peptide is not likely to span the entire
lipid bilayer (Brunner, 1989). Recent photolabeling studies have
demonstrated that also the intact virus interacts with liposomes
at the pH and temperature of fusion through HA2 solely (Brunner
et al., 1991; Stegmann et al., 1991).

Role of the Fusion Peptide: Because of the crucial importance of the HA2 N-terminus in the fusion of influenza virus several studies have focussed on the interaction of synthetic peptides corresponding to the fusion segment with lipid membranes (Lear and DeGrado, 1987; Murata et al., 1987; Wharton et al., 1988). Recently, we have extended this work to an investigation of the interaction of a number of peptides derived from the HA of the X-31 strain of the virus and several variants (Rafalski et al., 1991). The sequence changes in these variant peptides were derived from mutants of the Japan strain HA generated by site-directed mutagenesis (Gething et al., 1986). One striking mutant, in which the HA2 N-terminal glycine residue was replaced for a glutamic acid, did not show any significant fusion activity in a cell-cell fusion assay, after expression of the HA on the cell surface from a transfected gene (Gething et al., 1986).

In the synthetic-peptide system, the same replacement results in a substantial decrease of the peptide's ability to interact with lipid bilayer membranes (Rafalski et al., 1991). This is shown in Figure 6 , which presents the intrinsic Trp fluorescence of the "wild-type" peptide (WT) and that of the mutant (E1), in the absence and presence of PC large unilamellar vesicles (LUV) at pH 5.0. The blue shift and increase in intensity of the Trp fluorescence are significantly smaller for E1 than for the WT peptide, indicating that E1 penetrates less extensively into the hydrophobic interior of the liposomal bilayer. Remarkably, the WT peptide's ability to penetrate into liposomal bilayers is pH dependent, the shift of the fluorescence being marginal at pH 7.4 (not shown). If the fusion peptides in the HA trimer promote fusion by penetrating into the target membrane (see below), the pH dependence of this penetration proper might provide an additional level of pH control to the activation of the fusion capacity, besides the initial pH control at the level of the exposure of the peptides.

The WT peptide was also found to cause fast and extensive release of aqueous contents from PC LUV at pH 5.0, whereas at neutral pH release was negligible. On the other hand, at either pH value, the E1 peptide exhibits little capacity to disturb the

328

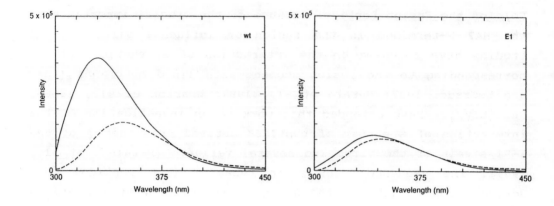

Figure 6: Intrinsic Trp fluorescence of HA-derived fusion peptides (1 μM) in the absence (dashed lines) and presence (solid lines) of PC LUV (100 μM lipid) at pH 5.0. The sequence of the WT peptide is GLFGAIAGFIENGWEGMIDG-amide (derived from the X-31 HA); in the E1 peptide the N-terminal Gly is replaced for a Glu. (Taken, with permission, from Rafalski et al., 1991).

permeability barrier of liposomes (Rafalski et al., 1991).

Circular dichroism studies revealed that it is the inability of the E1 peptide, compared to the WT, to adopt an α-helical structure upon interaction with lipids that is primarily responsible for precluding its penetration into lipid bilayer membranes (Rafalski et al., 1991).

Fusion Models: Several models for the mechanism of the HA-mediated membrane fusion reaction have been proposed (White et al., 1986; Stegmann et al., 1990; Ellens et al., 1990; White, 1990; Bentz et al., 1991). A central issue in this respect pertains to the precise role of the fusion peptide. It is attractive to postulate that the fusion peptides, once exposed as a result of the low-pH-induced conformational change in the HA, penetrate into the target membrane. However, this mechanism poses a problem: With the HA1 subunits protruding some 13.5 nm from the viral membrane, it is difficult to envisage how the fusion peptides, located at about 3.5 nm from the viral membrane, would bridge to remaining 10-nm gap to the target membrane. The "flower" model, originally proposed by White et al. (1986), on the basis of the observation that at low pH the

HA1 subunits dissociate (Doms and Helenius, 1986), suggests that the HA trimer structure opens up, the tops of the monomers folding back to the viral membrane, thus allowing the HA2 N-termini to reach the target membrane. However, in a recent study, Stegmann et al. (1990) provide evidence indicating that the final, low-pH, structure of the HA, in which the HA1 subunits are dissociated, is not fusion active. Rather, an intermediate low-pH conformation (White and Wilson, 1987), in which the trimer has not opened up yet, would exhibit fusion activity. The evidence is based on the observation that the virus has considerable and sustained fusion capacity at 0 °C, under which condition the tops of the HA monomers do not dissociate. Consistent with the notion that the open trimer conformation is not fusion-active, at 37 °C, when the trimer does open up, fusion activity is expressed only transiently, at least for the H3 subtype viruses (Stegmann et al., 1986, 1987b; Puri et al., 1990).

It appears that low pH induces a rapid exposure of the fusion peptides, even at 0 °C (White and Wilson, 1987; Stegmann et al., 1990); however, under these low-temperature conditions, fusion commences only after a considerable lag phase (Stegmann et al., 1990). Already early on in this lag phase the fusion peptides interact hydrophobically with the target membrane (Stegmann et al., 1991). It is proposed that the entire HA trimer tilts to allow the fusion peptides to reach the target membrane in the absence of dissociation of the HA1 subunits. According to the model, the continued lag, after penetration of the peptides into the target membrane until the onset of fusion, would involve the formation of an oligomeric fusion complex, consisting of several HA spikes, eventually resulting in the formation of a fusion pore (Spruce et al., 1989). There is evidence indicating the involvement of several HA spikes in the active fusion complex (Doms and Helenius, 1986; Ellens et al., 1990).

As an alternative to the tilting of largely intact HA trimers and penetration of the fusion peptides into the target membrane, it has been proposed that the HA trimers during the course of the fusion process remain in an upright position (Ellens et al.,

1990; White, 1990; Bentz et al., 1990). In this model the site of action of the fusion peptide is in between the apposed membranes, providing a hydrophobic surface for "wetting" of the inner aspect of a collar of HA trimers with lipid. The essential difference between this model and the models discussed above is that never in the course of the fusion process the two membranes achieve molecular contact; rather, the fusion peptides pull lipid into the gap between the membranes, thus forming an inter-membrane intermediate that eventually breaks into a fusion pore.

VIROSOME-MEDIATED INTRODUCTION OF FOREIGN MOLECULES INTO CELLS

Virosomes can be used as a vehicle for the introduction of foreign molecules into cells. As illustrated below, both water-soluble and membrane-associated compounds can be delivered.

Delivery of Water-Soluble Molecules: To explore the capacity of virosomes to deliver water-soluble molecules to the cytoplasm of cultured cells, we have used the A chain of Diphtheria toxin (DTA) as a marker. DTA is the toxic subunit of Diphtheria toxin; it very efficiently inhibits cellular protein synthesis by ADP-ribosylation of elongation factor 2 (for a review, see Pappenheimer, 1977). However, DTA when added to cells, is not toxic, as it can not bind to cellular receptors for the toxin, a capacity that in the intact toxin is located on the B subunit. We have encapsulated isolated DTA in virosomes during their preparation, and investigated whether this DTA is delivered to the cytoplasm of BHK-21 cells via endocytic uptake of the virosomes and fusion at the level of the endosomes (R. Bron, A. Ortiz and J. Wilschut, in preparation).

Figure 7 shows that DTA encapsulated in fusogenic virosomes induces a complete inhibition of the cellular protein synthesis. On the other hand, free DTA or empty virosomes have no effect, while furthermore the effect of virosome-encapsulated DTA can be blocked completely with NH_4Cl or by a pretreatment of the virosomes alone at low pH, causing an irreversible inactivation of

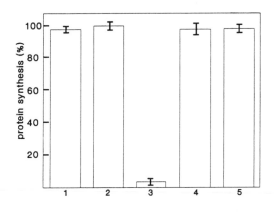
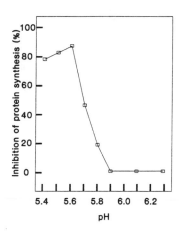

Figure 7 (left): Inhibition of protein synthesis in BHK-21 cells by DTA encapsulated in influenza (X-99) virosomes (at a concentration of approx. 5 DTA molecules per virosome). Virosomes were allowed to bind to cell monolayers in the cold. After washing, the cells were cultured for 6 h, after which protein synthesis was measured by incorporation of ^{3}H-leucine in TCA-precipitable material. 1, free DTA; 2, empty virosomes; 3, fusion-active DTA-virosomes; 4, fusion-inactivated DTA-virosomes; 5; fusion-active DTA-virosomes, but in the presence of 20 mM NH$_4$Cl.

Figure 8 (right): Fusion of DTA-containing virosomes with the plasma membrane of BHK-21 cells. After binding of the virosomes to the cell surface in the cold, the cells were treated for 2 min with buffers of the desired pH value at 37 $^{\circ}$C, washed with neutral-pH medium and processed as in Figure 7.

their fusion activity. Preliminary quantitation of the virosome-mediated cytoplasmic delivery of DTA, involving titration of the DTA-containing virosomes with empty virosomes indicates that, under the conditions employed, approx. 100-1000 virosomes per cell deliver their contents via fusion at the level of the endosomes, consistent with results obtained with the pyrene-PC assay (Figure 5). DTA delivery can also be accomplished through fusion of the virosomes at the cellular plasma membrane (Figure 8).

Insertion of Membrane-Associated Molecules: As an example of a biologically-active membrane-associated compound, we have studied bacterial endotoxic lipopolysaccharide (LPS). LPS is a potent activator of, among other cells, B-lymphocytes. The mechanism of action of LPS is not known, but it is believed that

332

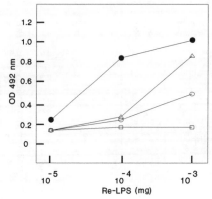

Figure 9: Activation of 70Z/3 pre-B-cells by free LPS and LPS coreconstituted in influenza virosomes (X-99). After binding of the virosomes (containing 100 μg of Re-LPS from S. minnesota R595 per μmol of virosomal phospholipid) at various concentrations to the cells at 8 °C for 60 min, the cells were washed and cultured for 24 h, after which kappa chain expression on the cell surface was measured with an ELISA-type assay (Ohno and Morrison, 1989). Solid circles, fusion-active LPS-virosomes; open circles, fusion-inactivated LPS-virosomes; triangles, free LPS; squares, control virosomes.

it acts at the level of the plasma membrane. When free LPS is added to cells, it, in part, partitions into the plasma membrane and/or interacts with a putative receptor, eventually resulting in cell activation. Figure 9 shows that LPS, coreconstituted in the membrane of virosomes that are subsequently fused to the plasma membrane of B-cells, is at least an order of magnitude more potent in activating the cells than free LPS (J. Dijkstra, A. de Haan, R. Bron, J. Wilschut and J.L. Ryan, submitted). Since fusion-inactivated LPS-virosomes induce little cell activation (Figure 9), the potentiation of the action of LPS is due to its fusion-mediated insertion into the cell plasma membrane.

PERSPECTIVES

The observation of influenza virus fusion at 0 °C, proceeding after a considerable lag period during which presumably an oligomeric fusion complex is formed (Stegmann et al., 1990), provides a promising basis for a further unravelling of the structural details of the fusion-active conformation of the hemagglutinin and its interaction with lipids. We believe that ultimate elucidation of the fusion mechanism will require the functional reconstitution of pure HA in a well-defined lipid environment, and investigation of such virosomes by the combined application of advanced biophysical, biochemical and morphological techniques. Besides, virosomes appear to provide an efficient carrier system for the introduction of foreign molecules into cells, with numerous potential applications in areas ranging from cell-biological research and membrane biotechnology to drug delivery and vaccine development.

Acknowledgements: We acknowledge the support from The Netherlands Organization for Scientific Research (fellowship to R.B.), the US National Institutes of Health (research grant AI25534), EMBO (long-term fellowship to A.O), The DuPont Company (grant to the University of Groningen), ImClone Systems, Inc., New York (grant to the University of Groningen) and Duphar B.V., Weesp, The Netherlands (ample supplies of virus) to our work.

REFERENCES

Bentz, J., Ellens, H. and Alford, D. (1990) FEBS Lett. 276, 1-5
Brunner, J. (1989) FEBS Lett. 257, 369-372
Brunner, J., Zugliani, C. and Mischler, R. (1991) Biochemistry 30, 2432-2438
Doms, R.W., Helenius, A. and White, J. (1985) J. Biol. Chem. 260, 2973-2981
Doms, R.W. and Helenius, A. (1986) J. Virol. 60, 833-839
Doms, R.W., White, J., Boulay, F. and Helenius, A. (1990) In: Membrane Fusion (Wilschut, J. and Hoekstra, D., Eds.), Marcel Dekker, Inc., New York, pp. 313-335
Ellens, H., Bentz, J., Mason, D., Zhang, F. and White, J. (1990) Biochemistry 29, 9697-9707
Gething, M.J., Doms, R.W., York, D. and White, J. (1986) J. Cell. Biol. 102, 11-23
Harter, C., Bächi, T., Semenza, G. and Brunner, J. (1988) Biochemistry 27, 1856-1864
Harter, C., James, P., Bächi, T., Semenza, G. and Brunner, J. (1989) J. Biol. Chem. 264, 6459-6464
Lear, J.D.and DeGrado, W.F. (1987) J. Biol. Chem. 262, 6500-6505
Marsh, M. and Helenius, A. (1989) Adv. Virus Res. 36, 107-151
Matlin, K.S., Reggio, H., Helenius, A. and Simons, K. (1981) J. Cell Biol. 91, 601-613
McClure, M.O., Marsh, M. and Weiss, R.A. (1988) EMBO J. 7, 513-518.
Mellman, I., Fuchs, R. and Helenius, A. (1986) Annu. Rev. Biochem. 55, 663-700

334

Murata, M., Sugahara, Y., Takahashi, S. and Ohnishi, S. (1987)
    J. Biochem. 102, 957-962
Ohno, N. and Morrison, D.C. (1989) Eur. J. Biochem. 186, 629-636
Pal, R., Barenholz, Y. and Wagner, R.R. (1988)
    Biochemistry 27, 30-36
Pappenheimer, A.M., Jr. (1977) Annu. Rev. Biochem. 46, 69-94
Puri, A., Booy, F., Doms, R.W., White, J.M. and Blumenthal, R.
    (1990) J. Virol. 64, 3824-3832
Rafalski, M., Ortiz, A., Rockwell, A., Van Ginkel, L.C., Lear,
    J.D., DeGrado, W.F. and Wilschut, J. (1991) Biochemistry 30,
    in press
Skehel, J.J., Bayley, P.M., Brown, E.B., Martin, S.R., Water-
    field, M.D., White, J.M., Wilson, I.A. and Wiley, D.C. (1982)
    Proc. Natl. Acad. Sci. USA 79, 968-972
Spruce, A.E., Iwata, A., White, J.M. and Almers, W. (1989)
    Nature 342, 555-558
Stegmann, T., Hoekstra, D., Scherphof, G. and Wilschut, J.
    (1985) Biochemistry 24, 3107-3113
Stegmann, T., Hoekstra, D., Scherphof, G. and Wilschut, J.
    (1986) J. Biol. Chem. 261, 10966-10969
Stegmann, T., Morselt, H.W.M., Scholma, J. and Wilschut, J.
    (1987a) Biochim. Biophys. Acta 904, 165-170
Stegmann, T., Booy, F.P. and Wilschut, J. (1987b)
    J. Biol. Chem. 262, 17744-17749
Stegmann, T., Morselt, H.W.M., Booy, F.P., Van Breemen, J.F.L.,
    Scherphof, G. and Wilschut, J. (1987c) EMBO J. 6, 2651-2659
Stegmann, T., Nir, S. and Wilschut, J. (1989a)
    Biochemistry 28, 1698-1704
Stegmann, T., Doms, R.W. and Helenius (1989b)
    Annu. Rev. Biophys. Biophys. Chem. 18, 187-211
Stegmann, T., White J. and Helenius, A. (1990)
    EMBO J. 9, 4231-4241
Stegmann, T., Delfino, J.M., Richards, F.M. and Helenius, A.
    (1991) J. Biol. Chem. 266, in press
Struck, D.K., Hoekstra, D. and Pagano, R.E. (1981)
    Biochemistry 20, 4093-4099
Wharton, S.A., Martin, S.R., Ruigrok, R.W.H., Skehel, J.J. and
    Wiley, D.C. (1988) J. Gen. Virol. 69, 1847-1857
White, J.M. (1990) Annu. Rev. Physiol. 52, 675-697
White, J.M. and Wilson, I.A. (1987) J. Cell Biol. 105, 2887-2896
White, J., Doms, R., Gething, M.J., Kielian, M. and Helenius, A.
    (1986) In: Virus Attachment and Entry into Cells (Crowell,
    T.R. and Longberg-Holmes, K., Eds.) Am. Soc. Microbiol.
    Washington DC, pp. 54-59
Wiley, D.C. and Skehel, J.J. (1987)
    Annu. Rev. Biochem. 56, 365-394
Wilschut, J. and Stegmann, T. (1988) In: Molecular Mechanisms of
    Membrane Fusion (Ohki, S. et al., Eds.) Plenum Press, New
    York, pp. 441-450
Wilschut, J., Scholma, J. and Stegmann, T. (1988) In: Biotechno-
    logical Applications of Lipid Microstructures (Gaber, B.P. et
    al., Eds.) Plenum Press, New York, pp. 105-126
Wilschut, J. and Hoekstra, D., Eds. (1990) Membrane Fusion,
    Marcel Dekker, Inc., New York

Index

B I R K H Ä U S E R
LIFE SCIENCES

Experientia Supplementum

Cell Motility Factors

Edited by
I. D. Goldberg
Long Island Jewish Medical Center, NY, USA

1991. 225 pages. Hardcover. ISBN 3-7643-2569-0 (EXS 59)

Contents:

Molecular Analysis of Amoeboid Chemotaxis: Parallel Observations in Amoeboid Phagocytes and Metastatic Tumor Cells – Adhesion Systems in Embryonic Epithelial-to-Mesenchyme Transformations and in Cancer Invasion and Metastasis – Neutrophil Chemotactic Factors – Purification and Characterization of Scatter Factor – Purification, Characterization and Mechanism of Action of Scatter Factor from Human Placenta – Scatter Factor Stimulates Migration of Vascular Endothelium and Capillary-Like Tube Formation – The Cellular Response to Factors which Induce Motility in Mammalian Cells – The Role of E-Cadherin and Scatter Factor in Tumor Invasion and Cell Motility – Heterogeneity Amongst Fibroblasts in the Production of Migration Stimulating Factor (MSF): Implications for Cancer Pathogenesis – Cell Motility, A Principal Requirement for Metastasis – Tumor Cell Autocrine Motility Factor Receptor – Interleukin-6 Enhances Motility of Breast Carcinoma Cells – Interleukin-6 Stimulates Motility of Vascular Endothelium – Computer Automation in Measurement and Analysis of Cell Motility *In Vitro*.

Contributors:

S.A. Aznavoorian, M.E. Beckner, J. Behrens, M.M. Bhargava, W. Birchmeier, B. Boyer, W. Carley, A. Coffer, J. Condeelis, M.A. Donovan, P.G. Dowrick, I. Ellis, U.H. Frixen, E. Gherardi, I.D. Goldberg, D. Grant, A.M. Grey, L. Harvath, R. Hofmann, A. Howell, S. Jaken, J.G. Jones, A. Joseph, H. Kleinman, Y. Li, L.A. Liotta, D. Liu, P.M. Luckett, I.R. Nabi, B. Palcic, M. Pendergast, M. Picardo, A. Raz, E.M. Rosen, G. Rushton, M. Sachs, E. Schiffmann, J.H. Schipper, A.M. Schor, S.L. Schor, J. Segall, P.B. Sehgal, E. Setter, S. Silletti, I. Spadinger, M.L. Stracke, I. Tamm, J.P. Thiery, G. Thurston, A.M. Valles, R.M. Warn, H. Watanabe, K.M. Weidner.

Please order through your bookseller or directly from:
Birkhäuser Verlag AG
P.O. Box 133
CH-4010 Basel / Switzerland

For orders originating from the USA or Canada:
Birkhäuser Boston Inc.
c/o Springer Verlag New York Inc.
44 Hartz Way
Secaucus, NJ 07096-2491 / USA

Birkhäuser

Birkhäuser Verlag AG
Basel · Boston · Berlin

BIRKHÄUSER
LIFE SCIENCES

Advances in Life Sciences

Methods
in Protein Sequence Analysis

Edited by
H. Jörnvall
J.-O. Höög
A.-M. Gustavsson
Karolinska Institute, Stockholm, Sweden

1991. 408 pages. Hardcover. ISBN 3-7643-2506-2 (ALS)

This book focuses on the methodological and interpretational aspects of protein analysis. Topics covered include novel approaches to sequencer instrumentation, different aspects of peptide purification, as well as new results in capillary electrophoresis, proteolysis, special chemical problems and modified residues. Mass spectrometry – including ion evaporation ionization and plasma desorption mass spectrometry – data bank comparisons and predictive methods, synergism with DNA analysis of three-dimensional structures, folding and interpretations are also treated. New possibilities are offered by emerging techniques such as C-terminal sequence analysis, the usefulness of capillary electrophoresis, mass spectrometry, more sophisticated sequencers and other instruments, as well as the impressive collection of structures available in data banks and valuable conclusions drawn from comparisons and functional correlations.

From the Contents:
– Sequencer methodology and instrumentation
– Sample preparation and analysis
– Modified residues, chemical problems and synthetic peptides
– Proteolysis
– Mass spectrometry
– Synergism with DNA analysis
– Predictions, data banks, patterns and tertiary structures

Please order through your bookseller or directly from:
Birkhäuser Verlag AG
P.O. Box 133
CH-4010 Basel / Switzerland

For orders originating from the USA or Canada:
Birkhäuser Boston Inc.
c/o Springer Verlag New York Inc.
44 Hartz Way
Secaucus, NJ 07096-2491 / USA

Birkhäuser

Birkhäuser Verlag AG
Basel · Boston · Berlin